AF566979

# ENGINEERING USABILITY

## FUNDAMENTALS, APPLICATIONS, HUMAN FACTORS, AND HUMAN ERROR

# ENGINEERING USABILITY

## FUNDAMENTALS, APPLICATIONS, HUMAN FACTORS, AND HUMAN ERROR

By

**B. S. Dhillon**
*Faculty of Engineering, University of Ottawa*
*Ottawa, Canada*

**AMERICAN SCIENTIFIC PUBLISHERS**
25650 North Lewis Way
Stevenson Ranch, California 91381-1439, USA

**AMERICAN SCIENTIFIC PUBLISHERS**
25650 North Lewis Way, Stevenson Ranch, California 91381-1439, USA
Tel.: (661) 254-0807, Fax: (61) 254-1207
E-mail: order@aspbs.com
WEB: www.aspbs.com

Engineering Usability: Fundamentals, Applications, Human Factors, and Human Error by B. S. Dhillon

This book is printed on acid free paper. ♾



**Library of Congress Catalog Number: 2004104510**
**International Standard Book Number: 1-58883-084-5**

PRINTED IN THE UNITED STATES OF AMERICA
10 9 8 7 6 5 4 3 2 1

## DEDICATION

*This book is affectionately dedicated to my ancestor, Ladha S. Dhillon, who founded my ancestral village of Pandori Ladha Singh and to all of his other descendants.*

# PREFACE

Today, billions of dollars are spent annually to produce new products using modern technologies. The usability of these products has become more important than ever before because of their increasing complexity, sophistication, and non-specialist users. For example, over 30% of all software development projects are cancelled before completion primarily because of inadequate user design input; resulting in a loss of $100 billion annually to the United States economy. Moreover, some studies indicate that 80% of product maintenance is due to unmet or unforeseen user requirements.

The term "Usability Engineering" was coined in the mid 1980's and it may simply be described as an approach to product development based on user or customer data and feedback. Over the years a large number of journal and conference proceedings publications directly or indirectly concerned with usability engineering have appeared. However, to the best of author's knowledge, there are only six or seven books that cover the subject in question in a significant depth. Moreover, none of these books cover the subjects of user errors, human reliability/error, and human factors upon which usability engineering is based. In addition, all of these volumes fail to cover the latest developments in the area of usability engineering. Currently, such information is either available in specialized articles or books, but not in a single volume.

This book is written to satisfy these vital needs. The sources of most of the material presented are given in the reference section at the end of each chapter. These will be useful to readers if they desire to delve deeper into a particular area. The topics covered in the volume are treated in such a manner that the reader will require no previous knowledge to understand the contents. At appropriate places, the book contains examples along with their solutions and at the end of each chapter there are numerous problems to test reader comprehension. A comprehensive list of 300 references on usability engineering is provided at the end of the book to give readers a view of the intensity of developments in the field.

The book is composed of twelve chapters. Chapter 1 presents various introductory aspects of usability engineering including need and importance of usability engineering, usability-related facts and figures, usability terms and definitions, and useful information on usability engineering. Chapters 2 and 3 are devoted to human factors basics for usability and human error and reliability basics, respectively. Some of the topics covered in Chapter 2 are classifications of human factors objectives, man-machine comparisons, typical human behaviors, human sensory capacities and body dimensions, and human factors data sources. Chapter 3 includes topics such as human error/reliability facts and figures, human tasks and their associated potential errors, operator stress characteristics, occupational stressors, and human reliability data sources and data.

Chapter 4 presents usability engineering life cycle stages and important associated areas. It covers topics such as basic features of design for usability, usability actions during system design phases, factors affecting usability within organizations, guidelines for managing the politics of usability, and usability benefits. A total of eleven usability evaluation methods are presented in Chapter 5. These are task analysis, cognitive walkthroughs, Kansei engineering, property checklists, expert appraisals, co-operative evaluation, cause and effect diagram, probability tree analysis, fault tree analysis, failure modes and effect analysis, and Markov method.

Chapter 6 presents various aspects of usability testing including usability testing goals, benefits, and limitations, types of usability tests, and usability test performance stages. Chapters 7 and 8 are devoted to user errors and usability costing, respectively. Some of the topics covered in Chapter 7 are user/operator error facts, figures, and examples, operator error causes, classification of user errors in human-computer interactive tasks, structured query language user errors, and methods for performing user error analysis. Chapter 8 includes topics such as usability costing-related facts and figures, usability engineering activities and costs, cost-benefit analysis of a usability study, cost of ignoring usability, cost effectiveness of usability evaluation methods, and models for estimating various types of usability engineering costs.

Chapter 9 presents seventeen mathematical models for usability assurance. Some of these models are movement time estimation model, character height estimation model, readability index models, channel capacity estimation model, control-display ratio estimation model, human performance estimation model, and lifting load estimation model. Chapter 10 is devoted to the topic of software usability. It includes topics such as need for considering usability during software development, software usability engineering process, software usability inspection methods, and software usability testing methods.

Chapters 11 and 12 are devoted to web usability and medical device usability, respectively. Some of the topics covered in Chapter 11 are common web design errors, web page design, navigation aids, web site design, and usability tools. Chapter 12 includes topics such as medical device users, use environments, user interfaces, a general approach for developing effective user interfaces of medical devices, guidelines for making medical device interfaces more user-friendly, and designing medical devices for old users.

This book will be useful to many people including design engineers, product users, human factors and usability specialists, computer-interface specialists, engineering administrators, computer system engineers, senior undergraduate students of engineering and psychology, and engineers-at-large.

The author is deeply indebted to Dr. S.N. Rayapati for seeding the idea to write this book.. The invisible inputs of the author's children, Jasmine and Mark, are also appreciated. Last, but not the least, I thank my wife, Rosy, for typing various portions of the book, help in proofreading, and tolerance.

**B. S. Dhillon**
Ottawa, Canada

# CONTENTS

# AUTHOR'S BIOGRAPHY

Dr. B. S. Dhillon is a Professor at the University of Ottawa. He has served as a Chairman/Director of Mechanical Engineering Department/Engineering Management Program for over 10 years at the same institution. He has published over 300 articles on human factors, human reliability, engineering management, reliability, safety, etc. He is or has been on the editorial boards of 7 international scientific journals. In addition, Dr. Dhillon has written 26 books on various aspects of human factors, engineering management, reliability, and design published by Wiley (1981), Van Nostrand (1982), Butterworth (1983), Marcel Dekker (1984), Pergamon (1986), etc. In 1986, he wrote first ever book on human reliability (published by the Pergamon Press) and it was translated into German and Chinese. Several of his other books have also been translated into languages such as Russian and Chinese.

He has served as General Chairman of two international conferences on reliability and quality control held in Los Angeles and Paris in 1987. Prof. Dhillon has served as a consultant to various organizations and bodies and has many years of experience in the industrial sector.

At the University of Ottawa, he has been teaching engineering management, reliability, safety, and related areas for over 23 years and he has also lectured in over 50 countries including keynote addresses at various scientific conferences held in North America, Europe, Asia, and Africa. Dr. Dhillon is recipient of many awards and a registered Professional Engineer in the Province of Ontario.

Professor Dhillon attended the University of Wales where he received a BS in electrical and electronic engineering and an MS in mechanical engineering. He received a Ph.D. in industrial engineering from the University of Windsor.

# Chapter 1

# INTRODUCTION

## CONTENTS



## 1. BACKGROUND

The emergence of usability engineering is deeply embedded in the discipline of human factors. The importance of usability/human factors in the design of engineering products may be traced back to 1901 in the Army Signal Corps contract document for the development of the Wright Brothers' Airplane; it clearly stated that the aircraft be "Simple to operate and maintain" [1].

However, human factors as a technical discipline emerged only after World War II, mainly due to the increasing complexity of military systems and the critical human role in operating them. Subsequently, all this knowledge was applied to other systems. Nonetheless, in 1946, a book entitled "Human Factors in Air Transport Design" consolidated the results of many isolated human factors-related projects and defined a comprehensive approach for human factors engineering for the entire aircraft [2].

In 1957, the Human Factors Society of America was incorporated and in 1976, the first North Atlantic Treaty Organization (NATO) Advanced Study Institute on "Man-Computer Interaction" was held [3]. In 1983, Association for Computing Machinery (ACM) Special Interest Group on Computer and Human Interaction (SIGCHI) was formed and its first conference on "Human Factors in Computing Systems" was held [4]. The term "Usability Engineering" was coined in the mid-1980s [5, 6].

Over the years a large number of publications directly or indirectly related to usability engineering have appeared. An extensive list of these publications is provided in the appendix.

## 2. NEED AND IMPORTANCE OF USABILITY ENGINEERING

Usability engineering is an effective approach to product development and is specifically based on customer feedback and data. Today, billions of dollars are being spent annually to produce new products using modern technologies. The usability of these products has become more important than ever before because of their increasing complexity, sophistication, and non-specialist users.

For example, over 30% of all software development projects are cancelled before completion primarily because of inadequate user design inputs; resulting in a loss of over $100 billion annually to the United

ISBN: 1-58883-084-5 / $35.00


*Engineering Usability: Fundamentals, Applications, Human Factors, and Human Error*
Authored by B. S. Dhillon

States economy. Moreover, some studies indicate that 80% of product maintenance is due to unmet or unforeseen user requirements.

All in all, it may be said that the key challenge in designing new products using modern technologies is how best to take advantage of potential users' skills in creating the most effective work environment; this may simply be called the usability engineering challenge.

## 3. FACTS AND FIGURES

This section presents various facts and figures directly or indirectly concerned with usability.

- It costs approximately $100 billion annually in lost productivity to American businesses because office workers "futz" with their machines an average of 5.1 hours per week [7].
- Over $500 billion is spent annually on items such as computers, networks, and information technology in the United States with a resulting loss in productivity [8].
- It is estimated that an average software program contains around 40 design flaws that impair workers' ability to use it effectively [8].
- User interface accounts for 47% - 60% of the lines of system or application code [9].
- A study reported that operator error accounts for over 50% of all technical medical equipment problems [10].
- A usability study conducted at Ford Motor Company revealed that relevant design changes resulted in the reduction of number of calls to the helpline from an average of three to none with the estimated savings of $100,000 [11].
- A study reported that 63% of all software projects overran their predicted costs, with the top four principal reasons all related to usability [12].
- 33%–50% reduction in the product development cycle has occurred with the application of usability engineering principles [13].
- A study conducted by American Airlines revealed that catching a usability-related problem early in the design could decrease the cost of rectifying it by 60%–90% [14–16].
- The Center for Devices and Radiological Health (CDRH) of the Food and Drug Administration (FDA) reported that approximately 60% of the deaths or serious injuries associated with medical devices were due to user error [17].
- A study reported that, in ergonomically Designed Office Equipment, absenteeism dropped from 4% to slightly greater than 1% [18].
- A study reported that because computers are often difficult to use, organizations frequently provide around $3,150 worth of technical support for each and every user of the equipment [19].
- Web usability is improving somewhere between 2% and 8% each year [20].
- A study revealed that the training time for new users of a standard personal computer was around 21 hours as opposed to only 11 hours for users of a more usable computer machine [21].
- A study reported that a usability engineering product achieved 80% higher revenues than the first release developed without considerations to usability engineering principles [15, 22].
- A study reported that usability-related design changes at International Business Machines (IBM) resulted in the reduction of 9.6 minutes per task; translating into a projected internal savings of $6.8 million for the entire organization for the year 1991 alone [23].
- It is estimated that approximately 80% of software maintenance cost is due to unforeseen/unmet user needs [24].

## 4. USABILITY TERMS AND DEFINITIONS

This section presents useful usability-related terms and definitions taken from published literature [6, 25–28].

- **Usability.** This is the quality of an interactive system with regard to factors such as user satisfaction, ease of use, and ease of learning.
- **User interface.** This is the physical representations and procedures for viewing and interacting with the system/product functionality.

- **Usability engineering.** This is iterative design and evaluation for providing customer feedback on the usefulness and usability of a system's/product's design and functionality throughout the development phase.
- **Usability evaluation.** This is any empirical or analytical activity directed at understanding or assessing the usability of an interactive product/system.
- **Human factors.** This is a body of scientific facts concerning the human characteristics (the term includes all psychosocial and biomedical considerations).
- **Human reliability.** This is the probability that a task will be accomplished successfully by a person at any required stage in system operation within a given minimum time (i.e., if the time requirement exists).
- **Human error.** This is the failure to perform a specified task (or the performance of a prohibited action), which could result in damage to property or disruption of scheduled operations.
- **User task.** This is a desired end-result of activities the product user would like to accomplish.
- **User-centred design.** This is an early and continuous involvement of users in the product design process.
- **User reaction survey.** This is a questionnaire completed by usability test participants during or after interaction with a specified product/system.
- **User background survey.** This is a questionnaire completed by usability test participants before interacting with the product/system.
- **Usability inspection.** This is an analytical approach in which usability specialists or experts evaluate the user interaction needed to carry out pivotal or crucial tasks with an interactive product/system to determine the problematic aspects of user input or system response.
- **Usability lab.** This is one or more rooms designed to simulate product/system use environment, normally instrumented with software and hardware, to determine behaviour of users as they perform their tasks.

## 5. USEFUL INFORMATION ON USABILITY ENGINEERING

This section presents selective and useful information directly or indirectly concerned with usability engineering, grouped under a number of classifications.

### 5.1 Books

- Bawa, J. Dorazio, P., Trenner, L., editors, The Usability Business: Making the Web Work, Springer-Verlag, New York, 2001.
- Mayhew, D. J., The Usability Engineering Lifecycle: A Practitioner's Handbook for User Interface Design, Morgan Kaufmann Publishers, San Francisco, 1999.
- Nielsen, J., Usability Engineering, Academic Press, Boston, 1993.
- Rosson, M. B., Carroll, J. M., Usability Engineering: Scenario-Based Development of Human-Computer Interaction, Academic Press, San Francisco, 2002.
- Rubin, J., Handbook of Usability Testing: How to Plan, Design, and Conduct Effective Tests, John Wiley and Sons, New York, 1994.
- Barnum, C. M., Usability Testing and Research, Longman, Inc., New York, 2002.
- Lindgaard, G., Usability Testing and System Evaluation: A Guide for Designing Useful Computer Systems, Chapman and Hall, London, 1994.
- Norlin, E., Usability Testing for Library Websites: A Hands-on Guide, American Library Association, Chicago, 2002.
- Adler, P. S., Winograd, T. A., editors, Usability: Turning Technologies Into Tools, Oxford University Press, New York, 1992.
- Nielsen, J., Mack, R. L., editors, Usability Inspection Methods, John Wiley and Sons, New York, 1994.
- Wiklund, M. E., Editor, Usability in Practice: How Companies Develop User-Friendly Products, AP Professional, Inc., Boston, 1994.
- Ghaoui, C., Editor, Usability Evaluation of Online Learning Programs, Information Science Publications, Inc., Hershey, Pennsylvania, 2003.
- Brink, T., Gergle, D., Wood, S. D., Designing Websites that Work: Usability for the Web, Morgan Kaufmann Publishers, San Francisco, 2002.
- Jordan, P. W., et al., editors, Usability Evaluation in Industry, Taylor and Francis, Inc., London, 1996.

### 5.2. Standards and Handbooks

- ISO 9241-11 (1998), Ergonomic Requirements for Office Work with Visual Display Terminals (VDTs): Guidance on Usability, International Organization for Standardization (ISO), Geneva, Switzerland.
- ISO 9241-13 (1998), Ergonomic Requirements for Office Work with Visual Display Terminals (VDTs): User Guidance, International Organization for Standardization (ISO), Geneva, Switzerland.
- ISO 13407 (1999), Human Centered Design Processes for Interactive Systems, International Organization for Standardization (ISO), Geneva, Switzerland.
- ETSI ETR 095, Human Factors: Guide for Usability Evaluations of Telecommunications Systems and Services, European Telecommunications Standards Institute (ETSI), Sophia Antipolis, France.
- ETSI ETR 116, Human Factors Guidelines for ISDN Terminal Equipment Design, European Telecommunications Standards Institute (ETSI), Sophia Antipolis, France.
- ETSI ETR 198, User Trials User Control Procedures in ISDN Video Telephony, European Telecommunications Standards Institute (ETSI), Sophia Antipolis, France.
- ETSI ETR 051, User Checklist for Telephones: Basic Requirements, European Telecommunications Standards Institute (ETSI), Sophia Antipolis, France.
- MIL-HDBK-761A, Human Engineering Guidelines for Management Information Systems, Department of Defense, Washington, D.C.
- MIL-STD-1908, Definitions of Human Factors Terms, Department of Defense, Washington, D.C.
- ISO TC 159/SC4, Ergonomics of Human-System Interaction, International Organization for Standardization (ISO), Geneva, Switzerland.
- MIL-STD-1478, Task Performance Analysis, Department of Defense, Washington, D.C.
- MIL-STD-1472D, Human Engineering Design Criteria for Military Systems, Equipment and Facilities, Department of Defense, Washington, D.C.
- MIL-HDBK-759B, Human Factors Engineering Design for Army Material, Department of Defense, Washington, D.C.

### 5.3. Journals and Magazines

- ACM Interactions
- User Modeling and User-Adapted Interaction (UMUAI)
- The SIGCHI Bulletin
- ACM Transactions on Computer-Human Interaction (TOCHI)
- International Journal of Human-Computer Studies
- Interacting with Computers
- Behaviour and Information Technology
- Human-Computer Interaction
- Human Factors
- Applied Ergonomics

### 5.4. Organizations

- Human Factors and Ergonomics Society, P.O. Box 1369, Santa Monica, California, USA.
- Usability Professional's Association, 140 N. Bloomingdale Road, Bloomingdale, Illinois, USA.
- Association for Computing Machinery (ACM) Special Interest Group on Computer and Human Interaction (SIGCHI), One Astor Plaza, 1515 Broadway, 17th Floor, New York, New York, USA.
- International Organization for Standardization (ISO), 1 rue de Varembe, Case postale 56 CH-1211, Geneva, Switzerland.

## 6. SCOPE OF THE BOOK

Each year, thousands of new products are designed, manufactured, and marketed using modern technologies around the world. The effectiveness of many of these products' usability is becoming increasingly important, in particular the ones that are concerned directly or indirectly with computers. Therefore, the knowledge of basic usability engineering principles and related areas is considered essential to professionals involved with the design, manufacture, and use of the new products.

This book is an attempt to meet this growing need. Prior knowledge of basic human factors, human error, and human reliability principles is not necessary to understand its contents because this material is covered in Chapters 2 and 3. This book will be useful to many professionals, including design engineers, product users, human factors and usability specialists, computer system engineers and computer-interface specialists, graduate and senior undergraduate students of engineering and psychology, computer scientists, and engineers-at-large.

## 7. PROBLEMS

1. Write an essay on the history of usability engineering.
2. Discuss the need for usability engineering.
3. List at least ten important usability-related facts and figures.
4. Define the following terms:
   - User interface
   - Human factors
   - User task
5. List five useful standards concerned with usability.
6. What is the difference between "usability" and "usability engineering"?
7. List at least five useful journals for obtaining the latest usability engineering-related information.
8. Discuss the relationships between usability and human factors.
9. What is usability inspection?
10. What is a usability laboratory?

## 8. REFERENCES


1. AMCP 706-133, Engineering Design Handbook: Maintainability Engineering Theory and Practice, Department of Defense, Washington, D.C., 1976.
2. McFarland, R. A., Human Factors in Air Transport Design, McGraw-Hill Book Company, New York, 1946.
3. Shackel, B., Richardson, S., Human Factors for Informatics Usability: Background and Overview, in Human Factors for Informatics Usability, edited by B. Shackel and S. Richardson, Cambridge University Press, Cambridge, U.K., 1991, pp. 1–19.
4. Concejero, P., Clarke, A., Carter, C., Muehlbach, L., Rushin, D., Kaasinen, E., Kolari, P., Chester, J., Current Available HF Guidelines and Standards, Report No. AC224/TID/2270/DR/P/002/b1, Prepared by Usability in Acts (USINACTS), October 1996. Available online at HTTP://www.hhi.de/USINACTS/publications.html.
5. Butler, K. A., Usability Engineering Turns Ten, Interactions, January 1996, pp. 59–75.
6. Rosson, M. B., Carroll, J. M., Usability Engineering: Scenario-Based Development of Human-Computer Interaction, Academic Press, San Francisco, 2002.
7. SBT Accounting Systems, 1997, Westlake Consulting Company, Inc., 5444 Westheimer, Unit 1510, Houston, Texas.
8. Landauer, T., The Trouble with Computers, Massachusetts Institute of Technology (MIT) Press, Boston, 1995.
9. Trenner, L., Bawa, J., Editors, The Politics of Usability: A Practical Guide to Designing Usable Systems in Industry, Springer-Verlag, London, 1998.
10. Dhillon, B. S., Reliability Technology in Health Care Systems, Proceedings of the IASTED International Symposium on Computers and Advanced Technology in Medicine, Health Care, and Bio-Engineering, 1990, pp. 84–87.
11. Kitsuse, A., Why aren't Computers ....., Across the Board, October 28, 1991, pp. 44–48.
12. Landauer, A. L., Prasad, J., Nine Management Guidelines for Better Cost Estimating, Communications of the ACM, Vol. 35, No. 2, 1992, pp. 51–59.
13. Bosert, J. L., Quality Function Deployment: A Practitioner's Approach, American Society for quality Control (ASQC) Quality Press, New York, 1991.
14. Chalupnik and Rinehart, 1992. Cited at HTTP://www.seast.usec.sun.com:80/usability/benefits.html.
15. Bevan, N., Cost Benefit Analysis, Report No. 3 version 1.1, September 8, 2000. Serco Usability Services, Alderney House, 4 Sandy Lane, Teddington, Middx, U.K.
16. Laplante, A., Put to the Test, Computer-World, Vol. 27, July 27, 1992, pp. 75–77.

17. Bogner, M. S., Medical Devices: A New Frontier for Human Factors, CSERIAC Gateway, Vol. IV, No. 1, 1993, pp. 12–14.
18. Schneider, M. F., Why Ergonomics Can No Longer Be Ignored, Office Administration and Automation, Vol. 46, No. 7, 1985, pp. 26–29.
19. Gibbs, W. W., Taking Computer to Task, Scientific American, No. 7, 1997, pp. 10–11.
20. Nielsen, J., PR on Websites: Increasing Usability, Alertbox, March 10, 2003. Available online at www.useit.com/alertbox/200303/0.html.
21. Nielson, J., Usability Engineering, Academic Press, Inc., Boston, 1993.
22. Wixon, D., Jones, S., Usability for Fun and Profit: A Case Study of the Redesign of the VAX RALLY, in Human-Computer Interface Design: Success Stories, Emerging Methods, and Real-World Context, edited by M. Rudisill, C. Lewis, P. G. Polson, T. McKay, Morgan Kaufmann Publishers, San Francisco, 1995.
23. Karat, C., Cost-Benefit Analysis of Usability Engineering Techniques, Proceedings of the Human Factors Society Conference, 1990, pp. 839–843.
24. Pressman, R. S., Software Engineering: A Practioner's Approach, McGraw Hill Book Company, New York, 1992.
25. Glossary of Terms used in Usability Engineering, Available online at http://www.ucc.ie/hfrg/baseline/glossary.html.
26. ISO 13407 (1999), User-Centered Design Process for Interactive Systems, International organization for Standardization (ISO), Geneva, Switzerland.
27. MIL-STD-721B (1986), Definitions of Effectiveness Terms for Reliability, Maintainability, Human Factors, and Safety, Department of Defense, Washington, D.C.
28. Meister, D., Human Factors in Reliability, in Reliability Handbook, edited by W. G. Ireson, McGraw Hill Book Company, New York, 1966, pp. 12.2–12.37.

# Chapter 2

# HUMAN FACTORS BASICS FOR ENGINEERING USABILITY

## CONTENTS



## 1. INTRODUCTION

The main reason for the existence of the discipline of human factors is that humans keep making errors while using machines. Otherwise, it would be rather difficult to justify the discipline's existence.

In modern times, the history of human factors may be traced back to 1898, when Frederick W. Taylor conducted various studies to determine the most effective design of shovels [1]. In 1911, Frank B. Gilbreth invented a scaffold that allowed bricklayers to conduct their activities at the most appropriate levels at all times. Consequently, the productivity of the bricklayers almost tripled (i.e., 120 to 360 bricks per hour). During the period between 1911 and 1945, many new developments took place in human factors. By 1945, human factors became a recognized discipline. In 1957, the Human Factors Society of America was incorporated and in 1992 it became known as the Human Factors and Ergonomics Society of America.

Over the years, a vast number of publications on the subject have appeared. In the publications terms, "human factors", "ergonomics", "human engineering", and "human factors engineering" have appeared interchangeably. Normally, the term "ergonomics" is less frequently used in U.S.A. and Canada than the other three aforementioned terms. It appears that the two most widely used terms throughout the world are "ergonomics" and "human factors". The former is derived from two Greek words: ergon (meaning work) and nomos (meaning law), and the latter is defined as a body of scientific facts concerning the characteristics of humans (the term embraces all psychosocial and biomedical considerations) [2–4].

This chapter presents some important aspects of human factors basics considered useful for engineering usability.

ISBN: 1-58883-084-5 / $35.00

## 2. CLASSIFICATIONS OF HUMAN FACTORS OBJECTIVES AND TYPES OF MAN-MACHINE SYSTEMS

Human factors has many objectives and they may be grouped under four classifications as follows [5]:

- **Classification I.** This includes basic operational objectives: improve system performance, reduce errors, and increase safety.
- **Classification II.** This includes objectives that affect operators and users: increase human comfort, improve the working environment, increase user acceptance, reduce fatigue and physical stress, increase ease of use, increase aesthetic appearance, and reduce boredom and monotony.
- **Classification III.** This includes objectives that have bearing on reliability, integrated logistic support, maintainability, and availability: increase reliability, reduce training requirements, reduce manpower requirements, and improve maintainability.
- **Classification IV.** This includes miscellaneous objectives such as: increase economy of production, and reduce losses of time and equipment.

There are basically three types of man-machine systems: manual systems, automated systems, and mechanical/semiautomatic systems [6]. Manual systems are composed of hand tools and other aids that are coupled together by the human operator controlling their operation. Usually, a craft worker is the human operator. Nonetheless, operators of these systems use their own energy as a power source, and they transmit and receive information from these systems during their use.

Automated systems perform all operational functions, including sensing, action, and information processing and decision making. These systems are required to be fully programmed to take appropriate measures under all contingencies as they occur. Three specific human functions associated with automated systems are monitoring, programming, and maintenance.

Mechanical/semiautomatic systems are composed of well-integrated physical components such as various types of powered machine tools and are usually designed to carry out their functions with minor variation. In these systems, typically the machine provides power and the human operator essentially looks after the controlling aspect.

## 3. MAN-MACHINE COMPARISONS AND TYPICAL HUMAN BEHAVIORS

During the design of engineering systems, decisions may have to be made whether to allocate certain functions to humans or to machines. In such circumstances, an effective knowledge of machines' and humans' capabilities/limitations is absolutely essential; otherwise, the correct decisions may not be made. Table 2.1 presents some comparisons of machines' and humans' capabilities and limitations [7].

Over the years, human factors researchers have conducted extensive research to predict human behaviour. They have successfully identified a great deal of typical human behaviour. Some of these typical behaviours, along with proposed design measures in parentheses, are as follows [7]:

- Humans tend to regard manufactured items as being safe (Design items/products in such a manner so that they cannot be used improperly. If this is not feasible, design in an appropriate mechanism for making users aware of all possible hazards).
- Humans have become very much accustomed to specific color meanings (Strictly observe current color-coding standards during design).
- Humans get easily confused with unfamiliar items (Do not design totally unfamiliar items to all potential users).
- Humans have a tendency to hurry (Design products/items in such a manner that takes into consideration the element of hurry by humans).
- Usually, humans know very little about their physical shortcomings (First learn effectively about human limitations and basic characteristics, and then develop design accordingly).
- Usually, humans use their hands first to explore or test (First of all, pay special attention to the handling aspect during product design. Otherwise, recommend strongly that product use requires a device supplied for eliminating the need to use hands).
- During loss of balance, humans instinctively reach for and grab the nearest object (Develop design in such a manner that it incorporates satisfactory emergency supports).

**Table 2.1** A comparison of machines' and humans' capabilities/limitations

| No. | Machine capability/limitation | Human capability/limitation |
|---|---|---|
| 1 | Free of social environment | Subject to social environment |
| 2 | Inflexible with respect to task performance | Relatively very flexible |
| 3 | Subject to ecological needs only | Subject to psychological, ecological, and physiological needs |
| 4 | Little or no induction capability | Highly capable of making inductive decisions in novel situations |
| 5 | Independent of g forces | Quite adversely affected by high g forces |
| 6 | Operation stops under overload conditions and usually fails at once | Quite capable of performing under transient overload (performance degrades gracefully) |
| 7 | Channel capacity can be expanded to satisfy the need | Limited channel capacity |
| 8 | Very expensive to have same memory capability as humans | Excellent memory |
| 9 | Designed strategy is executed all the time | Optimum strategy may not be followed all the time |
| 10 | Increase in complexity leads to serious maintenance problems | Relatively easy maintenance |
| 11 | Very poor in performing time-contingency analyses and predicting events in unfamiliar environments | Quite capable at this aspect |
| 12 | Subject to degradation in performance because of wear or lack of calibration | Subject to degradation in performance because of fatigue and boredom |
| 13 | Extremely useful to perform tasks such as data coding, amplification, or transformation | Quite unsuitable to perform such tasks |
| 14 | Limited in tolerance to factors such as vagueness, ambiguity, and uncertainty | A high degree of tolerance to such factors |
| 15 | Independent of interpersonal or other-related problems | Prone to stress as the result of such problems |
| 16 | Performance efficiency is unaffected by the anxiety factor | Performance efficiency is effected by the anxiety factor |
| 17 | Short-term memory can be expanded to any desirable and affordable level | Extremely limited short-term memory for factual matters |
| 18 | Free from factors such as motion sickness, coriolis effects, and disorientation | Prone to such factors |
| 19 | Usually, performs well only under ideal environments (i.e., noise-free, clean, etc.) | Quite capable to interpret an input signal even indistractive, noisy, and similar conditions |

- Humans expect that valve and faucet handles rotate counter-clockwise to increase the flow of gas, steam, or liquid (Ensure that such devices are designed according to human expectations).
- Humans expect electrically powered switches to move upward, to the right, etc., to activate power (Ensure that such devices are designed according to human expectations).
- The attention of the humans is drawn to items such as flashing lights, bright and vivid colors, loud noises, and bright lights (Ensure that stimuli of adequate intensity is designed in when attention requires stimulation).

## 4. HUMAN SENSORY CAPACITIES AND BODY DIMENSIONS

Humans possess many useful sensors: touch, hearing, smell, sight, and taste. A good understanding of these sensors can be useful to reduce various types of usability-related problems. This section discusses touch, noise, and sight in detail [8].

### Touch

This is closely related with humans' ability to interpret visual and auditory stimuli. The sensory cues received by the skin and muscles can be used to send messages to the brain. In turn, this relieves a part of the load from eyes and ears. This useful human quality can be used successfully in various engineering usability areas. For example, when a product user is expected to rely completely on his/her touch sensor, different knob shapes could be considered for use.

It may be added that the use of the touch sensor in various technical areas is not that new; it has been used for many centuries by craft workers for detecting surface roughness and irregularities in their work. In fact, past experiences indicate that the detection accuracy of surface irregularities improves dramatically when the individual moves an intermediate piece of paper or thin cloth over the object surface under consideration instead of just bare fingers [9].

#### Noise

This may simply be described as sounds that lack coherence, and human reaction to noise extends well beyond the auditory systems (i.e., to feelings such as fatigue, irritability, boredom, or well-being). Excessive noise can lead to various types of problems, including adverse effects on tasks requiring a high degree of muscular coordination and precision or intense concentration, reduction in workers' efficiency, and loss of hearing if exposed for long periods.

The human ear can detect sounds of frequencies from 20–20,000 Hz and is most sensitive to frequencies in the range of 600–900 Hz. Humans exposed to noise of frequencies between 4,000 and 6,000 Hz for long periods can suffer a major loss of hearing [8, 10].

#### Sight

This is stimulated by electromagnetic radiation of specific wavelengths, often referred to as the visible portion of the electromagnetic spectrum. The various parts of the spectrum, as seen by an individual's eye, appear to vary in brightness. According to studies, in daylight, the human eye is most sensitive to greenish-yellow light with a wavelength of approximately 5,500 Angstrom units [8]. Moreover, the eye sees differently from different angles.

Some important sight-related guidelines are as follows:

- Avoid relying on color as much as possible (where critical tasks may be carried out by fatigued persons).
- Select color in such a manner so that color-weak persons do not get confused.
- Aim to use red filters with wavelengths greater than 6,500 Angstrom units.

### 4.1. Body Dimensions

Information on body dimensions is very important to designers for allocating workspace and tasks to involved individuals, since most engineering products are operated and maintained by humans. Usually, requirements associated with the human body are outlined in product design specification. There are two basic sources for obtaining information on body dimensions: experiments and anthropometrics surveys.

Experiments simulate the conditions in question for the collection of required data. More specifically, first the conditions in question are simulated by experiments, and then the required data are collected. In the case of anthropometrics surveys, measurements of a sample of the population are taken. Usually, the results of these measurements are presented in the form of percentiles, ranges, and means.

Body measurements/dimensions are grouped under the following two main categories:

- **Dynamic.** In this case, measurements usually vary with body movements and include those taken with subjects in different working positions and arm and leg reaches.
- **Static.** In this case, measurements include everything ranging from dimensions of the largest aspects of body size to measurements of the distance between the pupils of the eyes.

Table 2.2 presents selected body dimensions of male and female U.S. adults (18 to 79 years) [6, 7, 11].

## 5. VIBRATIONS' EFFECTS ON HUMANS, EFFECTIVE ILLUMINATION LEVELS, AND GLARE AND GLARE REDUCTION

Vibration may simply be described as the alternating motion of a surface or body in regard to a reference point. Usually engineering products (e.g., vehicles, machines, and appliances) vibrate. Moreover, vibrations may be transmitted to product operators/users by direct mechanical path or through various flanking paths, and the vibrating objects can make the entire body or part of it vibrate.

**Table 2.2** Selected body dimensions (in inches) of male and female U.S. adults (18 to 79 years)

| | | Male | | Female | |
|---|---|---|---|---|---|
| **No.** | **Body feature** | **5th (percentile)** | **95th (percentile)** | **5th (percentile)** | **95th (percentile)** |
| 1 | Seated eye height | 28.7 | 33.5 | 27.4 | 31.0 |
| 2 | Seat breadth | 12.2 | 15.9 | 12.3 | 17.1 |
| 3 | Height, sitting erect | 33.2 | 38 | 30.9 | 35.7 |
| 4 | Height, standing | 63.8 | 72.8 | 59 | 67.1 |
| 5 | Knee height | 19.3 | 23.4 | 17.9 | 21.5 |

The human body's physical responses to vibrations are the results of complex and sophisticated interactions between body masses, damping, couplings, and elasticities in the low-frequency range (i.e., up to 50 Hz). Human body parts such as eyeballs, hands, the head, and shoulders have certain resonant frequencies that amplify the motions transmitted to them. These frequencies can be changed through measures such as cushioned seats, handholds, mountings, and foot restraints.

Vibrations can affect human performance in the following three ways [5]:

- By modifying perception,
- By blurring visual images of dials and indicators, and
- By affecting control movements.

Furthermore, excessive vibrations can be painful, unpleasant, and hazardous. Table 2.3 presents seven important selected activities affected by various vibration frequencies [5]. Light is necessary to see things. It can be used artistically to generate aesthetically pleasing or appealing effects in places such as offices, restaurants, homes, and museums. Ancient Greeks and Romans were probably the first people to recognize the artistic value of light [5]. They designed their buildings in such a manner that made the most effective use of natural light. Nonetheless, improperly used light can distract, annoy, affect performance, and so on. Needless to say, professionals working in the illumination area have made great strides. Table 2.4 presents effective illumination levels for performing various activities [5, 6, 12].

Glare could be an important factor in the effectiveness of engineering products' usability. It is generated by brightness within the field of vision that is significantly higher than the luminance to which human eye is adapted, causing a loss in visual performance and visibility, discomfort, or annoyance. There are basically two types of glare: direct glare and specular or reflected glare [6]. Direct glare is produced by light sources in the field of view, while specular or reflected glare is produced by high brightness reflections from glossy or polished surfaces that are reflected toward an individual (e.g., reflections of overhead lights in a computer terminal CRT screen [6, 13]).

Variations in the level of glare can be categorized into three distinct types as shown in Fig. 2.1.

Blinding glare, as its name implies, is extremely intense. Even after its removal for an appreciable length of time, no object can be seen. Disability glare decreases visibility and visual performance, and usually is accompanied by discomfort to a certain degree. Discomfort glare causes discomfort but it does not

**Table 2.3** Seven selected activities affected by various vibration frequencies

| No. | Activity description | Frequency range (expressed in Hz) |
|---|---|---|
| 1 | Speech | 1–20 |
| 2 | Reading text | 1–50 |
| 3 | Reading instruments | 6–11 |
| 4 | Depth perception | 25–60 |
| 5 | Tactile sensing | 30–300 |
| 6 | Tracking | 1–30 |
| 7 | Head movements | 6–8 |

**Table 2.4** Effective illumination levels for performing selected activities

| No. | Activity description | Illuminance range in lux |
|---|---|---|
| 1 | Visual tasks performed occasionally (working spaces) | 100–200 |
| 2 | Very special visual tasks of extremely low contrast and small size | 10,000–20,000 |
| 3 | Visual tasks of low contrast or very small size (e.g., reading hand written material in hard pencil on poor quality paper) | 1,000–2,000 |
| 4 | Visual tasks of medium contrast or small size (e.g., poorly printed or reproduced material) | 500–1,000 |
| 5 | Visual tasks of low contrast and very small size performed over a prolonged period | 2,000–5,000 |
| 6 | Visual tasks of high contrast or large size (e.g., reading printed material) | 200–500 |

necessarily interfere with visibility or visual performance. The calculation of discomfort glare ratings for specific lighting layouts takes into consideration various situational factors that affect visual comfort, including room shape and size, illumination level, number and location of luminaries, room surface reflectances, and luminaire type, size, and light distribution.

Over the years, various ways and means have been proposed to reduce different types of glare. Three methods are discussed below [6].

(i) **Direct glare from windows.** Some of the important measures to reduce direct glare from windows are as follows:
- Make use of only those windows that are set at a reasonable distance above the floor.
- Make use of blinds, shades, or louvers.
- Install an outdoor overhang above the window in question.
- Make use of light surrounds.

(ii) **Reflected glare.** Some of the important measures to reduce the reflected glare are as follows:
- Minimize luminaires' luminance level.
- Use light diffusing surfaces (e.g., no glossy paper and flat paint).
- Position light sources or work areas in such a way that the reflected light is not directed toward the eyes.
- Provide effective general illumination.

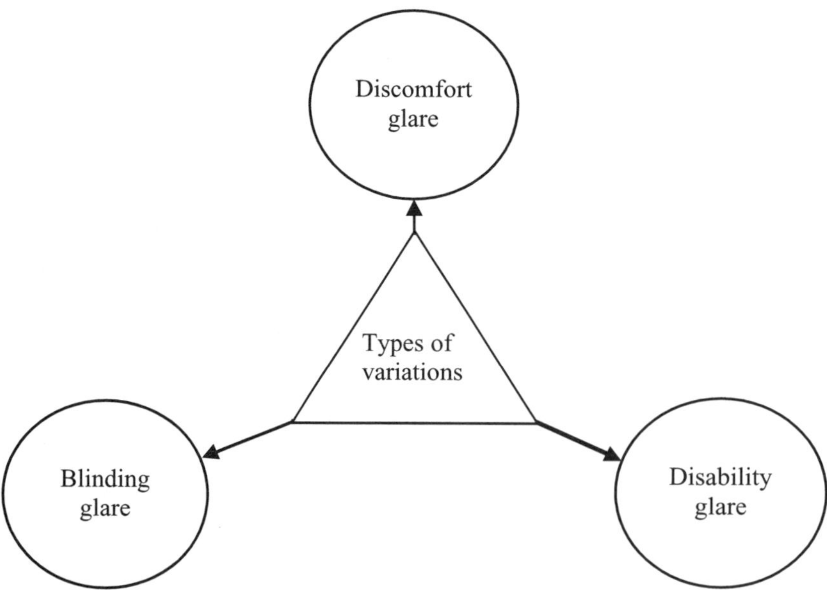


**Fig. 2.1** Types of variations in the level of glare

(iii) **Direct glare from luminaires.** Some of the important measures to reduce the direct glare from luminaires are as follows:
- Place luminaires as far from the line of sight as possible.
- Choose luminaires with low discomfort glare ratings.
- Employ hoods, visors, and light shields when it is impossible to reduce glare sources.
- Reduce the light sources' luminance.
- Increase the luminance of the area around the glare source so that the luminance (brightness) ratio is at minimum.

## 6. WORK LOAD AND WORK LOAD INDEX

Work load is based on two assumptions: (i) tasks are carried out with respect to time, (ii) the human operator possesses limited capacity for processing information within the time available [14–15]. Overload occurs when the events per unit time exceed the power of operator to choose and process them in an effective manner. Past experiences indicate that an overhead operator will make errors and omit responses because he/she has too many things to do within the time available.

Although, overload occurs frequently in complex tasks with multiple activities, it can also occur with a single activity. For example, a difficult single task can overload an individual's capacities because of the required intense mental effort.

There are no absolute workload measures. In fact, the current measures are relative and useful only for comparing two or more situations. For example, work load under normal and emergency situations or work load with two completely different displays and controls.

One useful work load measure or index is defined as follows [16]:

$$WLI = \frac{(T_r)(100)}{T_a} \tag{2.1}$$

where

$WLI$ is the work load index.
$T_a$ is the time available.
$T_r$ is the time required.

Although this index is quite useful, its three principal shortcomings are the work load inherent in continuous control tasks is not taken into consideration, the cognitive activity is not considered, and the work load in doing two or more things simultaneously is not considered [14].

## 7. HUMAN FACTORS DATA SOURCES

Data play an instrumental role in various types of human factors analysis. These data include body dimensions and weights, human error rates, energy expenditure per grade of work, permissible noise exposure per unit time, etc. Some of the forms in which these data may exist are as follows:

- Quantitative data tables.
- Mathematical functions and expressions.
- Expert judgments.
- Graphic representations.
- Design standards.

Nonetheless, some of the important sources for collecting human factors-related data are test reports, user experience reports, published standards, previous experience, published literature, and product development phase [17]. A list of useful documents/publications for obtaining human factors-related data is presented in Table 2.5.

**Table 2.5** Useful documents for obtaining human factors-related data

| No. | Document/publication |
|---|---|
| 1 | Woodson, W.E., Human Factors Design Handbook, McGraw-Hill Book Company, New York, 1981. |
| 2 | Lighting Handbook, Prepared by the Illumination Engineering Society, New York, 1971. |
| 3 | White, R.M., The Anthropometry of United States Army Men and Women: 1946–1977, Human Factors, Vol. 21, 1979, pp. 473–482. |
| 4 | Anthropometry for Designers, Anthropometrics Source Book 1, Report No. 1024, National Aeronautics and Space Administration (NASA), Washington, D.C., 1978. |
| 5 | Meister, D., Sullivan, D., Guide to Human Engineering Design for Visual Displays, Report No. AD 693-237, 1969. Available from the National Technical Information Service (NTIS), Springfield, Virginia, USA. |
| 6 | Parker, J.F., West, W.R., editors, Bioastronautics Data Book, Report No. NASA-SP-3006, U.S. Printing Office, Washington, D.C. |
| 7 | DOD-HDBK-743A, Anthropometry of U.S. Military Personnel, Department of Defense, Washington, D.C. |
| 8 | MIL-STD-1472D, Human Engineering Design Criteria for Military Systems, Equipment, and Facilities, Department of Defense, Washington, D.C. |

## 8. PROBLEMS

1. Write an essay on the historical developments in human factors.
2. Define the following two terms:
   - Human factors
   - Ergonomics
3. Discuss the four classifications of human factors objectives.
4. Describe the following types of man-machine systems:
   - Manual systems
   - Automated systems
   - Mechanical/semi-automatic systems
5. Compare machines' and humans' capabilities/limitations.
6. List at least seven typical human behaviours.
7. Discuss human sensory capacities.
8. What is the difference between static and dynamic body measurements/dimensions?
9. Describe the following terms:
   - Discomfort glare
   - Blinding glare
   - Disability glare
10. List at least five good sources for obtaining human factors-related data.

## 9. REFERENCES


1. Dale Huchingson, R., New Horizons for Human Factors in Design, McGraw-Hill Book Company, New York, 1981.
2. MIL-STD-721B, Definitions of Effectiveness Terms for Reliability, Maintainability, Human Factors and Safety, August 1966. Available from the Naval Publications and Forms Center, 5801 Tabor Avenue, Philadelphia, Pennsylvania.
3. Meister, D., Human Factors in Reliability, in Reliability Handbook, edited by W. G., Ireson, McGraw-Hill Book Company, New York, 1966, pp. 12.2–12.37.
4. Dhillon, B. S., Human Reliability: with Human Factors, Pergamon Press, Inc., New York, 1986.
5. Chapanis, A., Human Factors in Systems Engineering, John Wiley and Sons, New York, 1996.
6. McCormick, E. J., Sanders, M. S., Human Factors in Engineering and Design, McGraw-Hill Book Company, New York, 1982.

7. Woodson, W. E., Human Factors Design Handbook, McGraw-Hill Book Company, New York, 1981.
8. AMCP 706-134, Engineering Design Handbook: Maintainability Guide for Design, Prepared by the United States Army Material Command, 5001 Eisenhower Avenue, Alexandria, Virginia, 1972.
9. Lederman, S., Heightening Tactile Impression of Surface Texture, in Active Touch, edited by G. Gordon, Pergamon Press, New York, 1978, pp. 40–45.
10. AMCP 706-133, Engineering Design Handbook: Maintainability Engineering Theory and Practice, Prepared by the United States Army Material Command, 50001 Eisenhower Avenue, Alexandria, Virginia, 1976.
11. MIL-STD-1472, Human Engineering Design Criteria for Military Systems, Equipment, and Facilities, Department of Defense, Washington, D.C.
12. RQQ Report No. 6, Selection of Illuminance Values for Interior Lighting Design, Journal of Illuminating Engineering Society, Vol. 9, No. 3, 1980, pp. 188–190.
13. Hultgren, G., Knave, B., Discomfort Glare and Disturbances from Light Reflections in an Office Landscape with CRT Display Terminals, Applied Ergonomics, No. 1, Vol. 5, 1974, pp. 2–8.
14. Adams, J. A., Human Factors Engineering, Macmillan Publishing Company, New York, 1989.
15. Lane, D. M., Limited Capacity, Attention Allocation, and Productivity, In Human Performance and Productivity, edited by W. C. Howell and E. A. Fleishman, Lawrence Erlbaum Associates, Inc., Hillsdale, New Jersey, 1982, pp. 121–156.
16. Stone, G., Gulick, R. K., Gabriel, R. F., Use of Task-Timeline Analysis to Assess Crew Workload, In the Practical Assessment of Workload, edited by A.H. Roscoe, AGARDOgraph 282, Advisory Group for Aerospace Research and Development (AGARD), North Atlantic Treaty Organization, Brussels, Belgium, 1987, pp. 15–31.
17. Peters, G. A., Adams, B. B., Three Criteria for Readable Panel Markings, Product Engineering, Vol. 30, No. 21, 1959, pp. 55–57.

# Chapter 3

## HUMAN ERROR AND RELIABILITY BASICS

### CONTENTS



## 1. INTRODUCTION

Today, increasing attention is being paid to human error and reliability, because past experiences indicate that the human element accounts for a substantial proportion of system/product failures. Although human error and human reliability may convey basically the same thing to many people, at times, the proper understanding of their distinction could be quite important.

The main distinction between the two is demonstrated by their definitions. Human error is defined as a failure to perform a specified task (or the performance of a prohibited action), which could result in damage to property or disruption of scheduled operations [1, 2]. Similarly, human reliability is defined as the probability that a task will be accomplished successfully by a person at any required stage in system operation within a given minimum time (i.e., if the time requirement exists) [2–4].

The history of human error may be traced back to 1848 when the first anaesthetic death occurred [5]. More specifically, the death is associated with the occurrence of human error. However, human error/reliability has received serious attention only after the mid-1950s. In 1958, H.L. Williams pointed out that human-element reliability must be taken into consideration in the total system-reliability prediction; otherwise, the predicted reliability figure would not represent the true picture [6]. In 1960, two separate studies reported that human error was responsible for a large proportion of equipment failures [7, 8].

In 1973, a well-known journal on reliability published a special issue devoted to the subject of human reliability [9], and the first commercially available book on human reliability appeared in 1986 [4]. Over the years, a vast number of publications in the form of journal and conference proceeding articles, books, and technical reports on human error/reliability have appeared [10–12]. This chapter presents various aspects of human error/reliability basics considered useful for engineering usability.

ISBN: 1-58883-084-5 / $35.00

## 2. HUMAN ERROR/RELIABILITY FACTS AND FIGURES

This section presents various facts and figures directly or indirectly concerned with human error/reliability.

- A study of 23,000 defects in the manufacturing of parts for use in the nuclear area reported that around 82% of the defects were the result of human error [13].
- A study of 600 failure reports of rocket engines revealed that 35% of the malfunctions were human-initiated [14–15].
- A study of ship collisions, floodings, and groundings conducted over a period of four years revealed that 63.6% of these events were due to human error [16].
- Each year, around 100,000 Americans die due to various types of medical human errors and their estimated cost to the U.S. economy is between $17 billion and $29 billion [17].
- A study of 213 Air Defence System failures revealed that 25.8% were partially or wholly due to maintenance error [18].
- A study of 135 vessel failures occurring over the period from 1926–1988 reported that 24.5% were directly caused by humans [19].
- Over 50% of all technical medical equipment problems are caused by operators [20].
- A study of 552 aircraft (B-52) failures discovered that 36% were due to humans [15].
- An examination of 14 Australian studies published during the period from 1988–1996 revealed that 2.4%–3.6% of all hospital admissions were drug-related and 32%–69% were preventable [21].
- A study of tasks such as adjust, align, and remove performed by humans reported an average human reliability of 0.9871 [22].
- A study of 47 major aircraft accidents revealed that 51% were due to human error [15].
- Of all medical device-related deaths or injuries reported through the Food and Drug Administration's (FDA's) Center for Devices and Radiological Health (CDRH), 60% were attributed to user error [23].
- The annual cost of medication errors in the United States is estimated to be over $7 billion [24].
- A major teaching hospital in Hong Kong administered 16,000 anaesthetics in one year and reported 125 related critical incidents; in 80% of these incidents, human error was an important factor [25].
- In 1979, at O'Hare airport in Chicago, in a DC-10 aircraft accident, 272 persons died because repairmen followed improper maintenance procedures [26].
- A study of maintenance operations among commercial airlines discovered that 40–50% of the time, parts removed for repair were not defective [26].
- Over 90% of the documented air traffic control system errors were due to human operator [27].
- Up to 90% of accidents, both generally and in medical devices, are due to human error [28].
- A study of 35,000 missile failure reports revealed that 20–30% of the malfunctions were human-initiated [1].
- A study of "Defense" system failures conducted over a period of 4.5 years revealed that 40% of the malfunctions were caused by human error [29].
- A study of missile accidents conducted over a three-year period reported that 65% of the accidents were the result of human error [1, 30].
- Sixty percent of aircraft accidents are caused by human error [31].
- A study of 3,829 missile failure reports revealed that 20–53% of the malfunctions were due to human error [7].
- A study of 1,642 aircraft equipment failure reports revealed that 37% of the failures were due to human error [15].

## 3. HUMAN ERROR IN SYSTEM DEVELOPMENT STAGES, ITS CAUSAL FACTORS, AND CONSEQUENCES AND PROPORTIONAL CONTRIBUTION OF ASSEMBLY, OPERATOR, INSTALLATION, AND MAINTENANCE ERRORS DURING SYSTEM USEFUL LIFE

System development may be divided into four stages as shown in Fig. 3.1 [15].

The type of human error that occurs during the design stage is known as design error. Its three main causal factors are poor human engineering design, inappropriate function allocation, and failure to implement requirements. The main consequence of design error is inadequately designed equipment. The types of

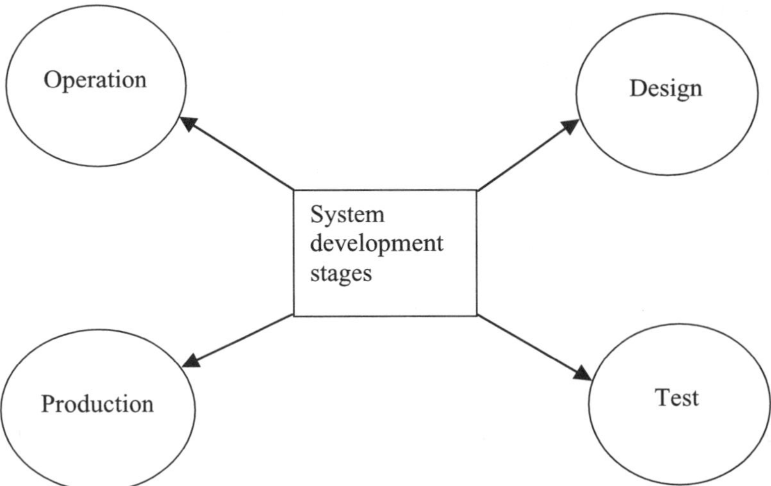


**Fig. 3.1** System development stages

human errors that occur during the production stage are known as fabrication error and inspection error. Some of the causal factors for the occurrence of these two types of errors are inadequate tools, wrong blueprints, poor workplace layout, poor human engineering design of equipment, inadequate skill or training, poor environment, and wrong instructions [15]. There are various consequences for the occurrence of fabrication and inspection errors, including higher cost, scrapped equipment, production delays, rejection of good equipment, and classification of malfunctioning equipment as functioning equipment.

The types of human errors that occur during the test stage are installation error, operating error, and maintenance error. The causal factors for the occurrence of these three types of errors include poor logistics, inadequate/incomplete technical data, poor workplace layout, and poor human engineering design. Some of the consequences for the occurrence of these errors are delay in system operations, system breakdown, failure to accomplish test, human-initiated malfunctions, degradation in system performance, and possible danger and loss of life.

The types of human errors that occur during the operation stage are as follows:

- Maintenance error
- Installation error
- Operating error

Some of the causal factors for these three types of errors are task complexity, inadequate logistics, overload conditions, inadequate/incomplete technical data, inadequate motivation, inadequate training/skill, poor workplace layout, poor human engineering design, and inadequate environment. The consequences for the occurrence of these errors include failure to accomplish mission, system breakdown, delay in system operations, possible danger and loss of life, human-initiated failures, and degradation in system performance [15].

The proportional contribution of the different species of human error to system failure varies quite significantly during the system's useful life (i.e., from acceptance to the start of phase out). Fig. 3.2 shows the approximate contributions of assembly, installation, operator, and maintenance errors to system failure [15, 32].

## 4. HUMAN TASKS AND THEIR ASSOCIATED POTENTIAL ERRORS

Most of the human tasks or functions may be classified under the following five distinct areas [15, 33]:

- Sequencing
- Estimating and tracking
- Sensing, detecting, identifying, coding, and classifying
- Problem solving
- Decision making

There are various types of potential human errors associated with each of these areas. Sequencing-related potential human errors include omitting a procedural step, inserting an unnecessary procedural step, mis-

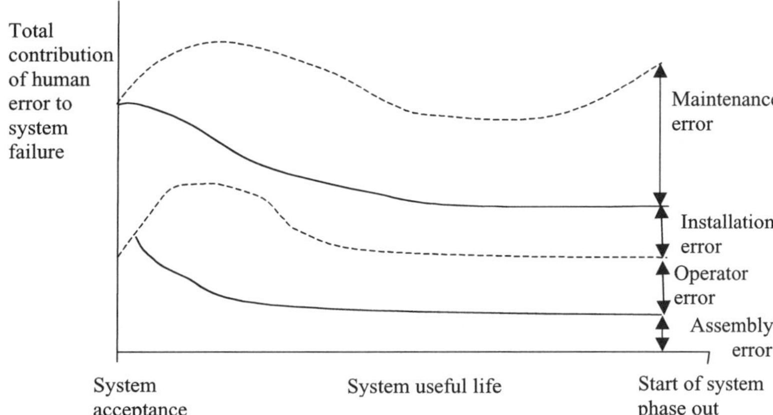


**Fig. 3.2** Approximate contributions of assembly, installation, operator, and maintenance errors to system failure

ordering procedural steps, and making a below-standard response. Estimating/tracking-related potential human errors include premature response to a target change, incorrect direction of control action, excessive continuance of control action, failure to respond to an obvious target change, inadequate continuance of control action, late response to a target change, excessive magnitude of control action, and inadequate magnitude of control action.

The sensing, detecting, identifying, coding, and classifying-related potential human errors include failure to record or report a signal change, recording or reporting a signal change in the incorrect direction, failure to record or report the appearance of a target, failure to monitor, assignment of a target to the wrong class, recording or reporting a signal change when none has occurred, and recording or reporting a target when none is in the field.

The problem solving-related potential human errors include failure to use available information to derive needed solution, acceptance of inadequate solution as final, and formulating erroneous rules or guiding principles. The decision making-related potential human errors include failure to apply an available rule, application of a fallacious rule, making an unnecessary or premature decision, wrong weighting of responses to a contingency, delaying a decision beyond the time it is required, failure to obtain or apply all relevant decision information, application of a correct but inappropriate rule, and failure to identify all reasonable alternatives [15, 33].

## 5. HUMAN PERFORMANCE EFFECTIVENESS VERSUS STRESS AND OPERATOR STRESS CHARACTERISTICS

The effectiveness of human performance is very much dependent upon the degree of stress. Past experiences indicate that over-stressed persons performing various types of tasks make more mistakes than persons working under normal stress. Many researchers have studied the relationship between human performance effectiveness and stress or anxiety over the past few years. The result of their efforts is depicted by Fig. 3.3 [34].

The figure shows that a moderate degree of stress is necessary to achieve optimal human performance effectiveness. It means, at very low stress, the task will be dull and unchallenging, thus human performance effectiveness will not be at its peak. In contrast, stress above a moderate level will result in decline in human performance. Some of the reasons for the decline are fear, worry, or other kinds of psychological stress.

Fig. 3.3 shows the human performance effectiveness curve divided into two regions, i.e., Regions I and II. In region I, human performance effectiveness increases with the increasing values of stress, and in region II, it decreases with the increasing values of stress. All in all, it may be concluded that a moderate level of stress is essential for achieving optimal human performance.

Usually, human operators face certain limitations in performing their assigned tasks. The probability of error occurrence increases dramatically when these limitations are violated. This probability can be reduced quite significantly by considering operator limitations or stress characteristics during the system/product design phase. Some of these stress characteristics are requirements for prolonged monitoring, having a very short decision-making time, ineffective feedback information to determine the correctness of actions taken,

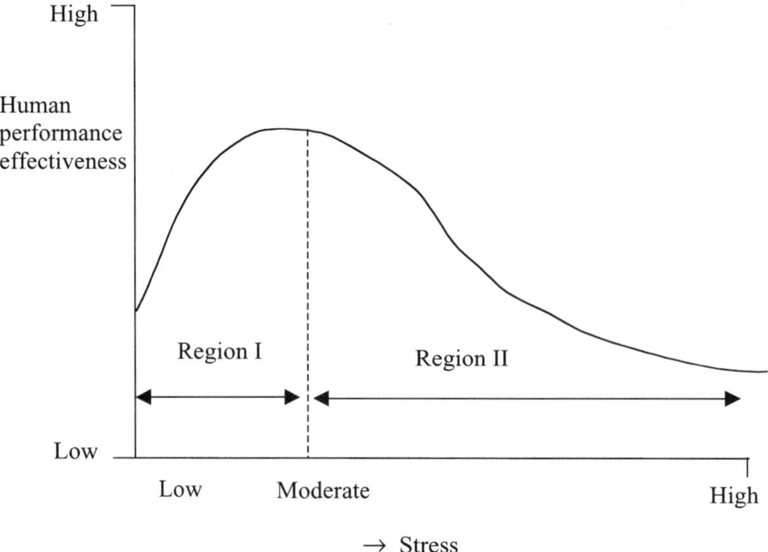


**Fig. 3.3** Human performance effectiveness versus stress curve

making a quick comparison of two or more displays, difficulty in discriminating more than one display, requirements to operate two or more controls simultaneously at high speed, the performance of tasks requiring a very long sequence of steps, requirements for making decisions on the basis of data obtained from various different sources, and requirements to perform task steps at high speed [35].

## 6. OCCUPATIONAL STRESSORS AND GENERAL STRESS FACTORS

There are many occupational stressors that can affect a person's reliability. They may be grouped under four distinct categories: occupational change-related, workload-related, occupational frustration-related, and miscellaneous [34].

The occupational change-related stressors are associated with factors that disrupt the person's cognitive, behavioural, and physiological patterns of functioning. Usually, stressors such as these are found in establishments concerned with productivity and growth. Some examples of occupational change are organizational restructuring, promotion, relocation, and scientific developments.

Workload-related stressors are associated with work under-load or overload. In the case of work under-load, the current tasks being carried out by the person provide insufficient stimulation. Some examples of work under-load are repetitive tasks, lack of applying any intellectual input, and lack of opportunity to use the person's acquired skills and expertise. On the other hand, in the case of work overload, the requirements of the job exceed the individual's ability to satisfy them in an effective manner.

Occupational frustration-related stressors are associated with problems of occupational frustration (i.e., conditions where the job inhibits the achievement of set goals or objectives in an effective manner). Examples of occupational frustration-related problems include poor career development guidance, ambiguity of role, bureaucracy difficulties, and lack of proper communication. Miscellaneous stressors include all other stressors than those comprising the previous three categories. Some examples of these stressors are as follows [34]:

- Excessive noise
- Poor interpersonal relationships
- Too little or too much lighting

There are many general factors that increase stress on humans and, in turn, affect an individual's performance/reliability. Some of these factors are as follows [34]:

- Working under extremely tight time schedules.
- Lacking the proper skill and experience to perform the ongoing job.

- Possibility of redundancy at work.
- Poor health.
- Excessive demands from superiors at work.
- Experiencing difficulties with spouse or children, or both.
- Poor chances for promotion
- Serious personal financial difficulties.
- Having to work with individuals with unpredictable temperaments.
- Unhappiness with the present job.

## 7. A METHOD FOR PREDICTING HUMAN RELIABILITY

Over the years, many methods have been developed to predict human reliability [4]. One of these methods is presented in this section. The method is known as Pontecorvo and is concerned with obtaining reliability estimates of discrete and separate subtasks with no accurate reliability values. These estimates are combined to obtain an overall estimate for tasks reliability. The method is quite useful to quantitatively assess the interaction of humans and machines during the initial design phase and for determining the performance of an individual acting alone.

The Pontecorvo approach is composed of six steps as shown in Fig. 3.4 [36]. Step 1 is concerned with the identification of tasks to be carried out. These tasks are usually identified at a gross level. More clearly, each task represents one complete operation. Step 2 is concerned with the identification of subtasks associated with each task. More specifically, those subtasks that are essential for completing each of these tasks.

Step 3 is concerned with gathering all relevant empirical performance data subject to environments under which the subtasks are to be carried out. Step 4 is concerned with rating each subtask with respect to difficulty level or potential for error. A 10-point scale is used in estimating the subtask rate (e.g., 10 for the most likelihood of error and 1 for the least likelihood of error). Step 5 is basically concerned with predicting subtask reliability by expressing the empirical data and judged data ratings in the form of a straight line, and testing it (i.e., the line) for goodness of fit. Step 6 is concerned with determining the task reliability by multiplying the subtask reliabilities.

These six steps are used to estimate the performance reliability of a person acting alone. By having a backup person, even for a certain amount of time, the probability of the task being performed correctly can be improved.

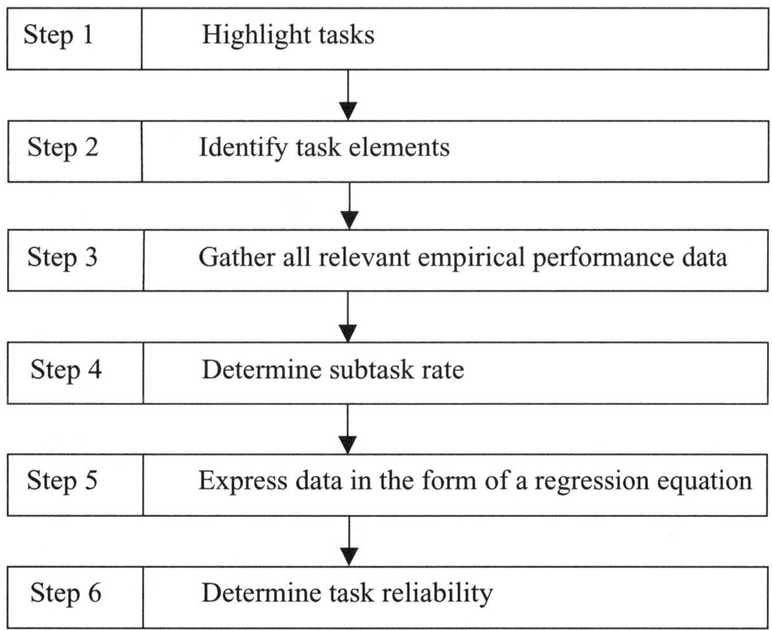


**Fig. 3.4** Pontecorvo method steps

Under such conditions, the overall reliability of two persons working together to accomplish a given task can be estimated by using the following equation [36]:

$$R_{otp} = \left[\left\{1-\left(1-R_{sp}\right)^2\right\}\alpha_{ba} + R_{sp}\alpha_{bu}\right] / \left(\alpha_{ba} + \alpha_{bu}\right) \quad \textbf{(3.1)}$$

where

$R_{otp}$ is the overall reliability of two persons working together to accomplish a given task.
$\alpha_{ba}$ is the percentage of time the backup person is available.
$\alpha_{bu}$ is the percentage of time the backup person is unavailable.
$R_{sp}$ is the single person's reliability.

---

**Example 3.1**

Two persons are working together to operate a sophisticated piece of equipment. The reliability of each person is 0.95 and the backup person is available only 80% of the time. More specifically, 20% of the time, the equipment is operated by a single individual. Calculate the probability of operating the equipment correctly.

By substituting the specified data values into Equation (3.1), we get

$$\begin{aligned} R_{otp} &= \left[\left\{1-(1-0.95)^2\right\}(0.80) + (0.95)(0.20)\right] / (0.80+0.20) \\ &= 0.988 \end{aligned}$$

It means the probability of operating the equipment correctly is 0.988.

---

## 8. HUMAN RELIABILITY DATA SOURCES AND DATA

Over the past 40 years, various types of human reliability data have been accumulated. Nonetheless, human reliability/error data can be collected through various different means, including published literature, experimental studies, self-made error reports, expert judgements, and human data recorder [37–38].

There are many specific sources or banks for obtaining human error/reliability data [39–40]. Some of these are as follows [41]:

- Aviation Safety Reporting System [42]
- Data Store [43]
- Human Reliability and Safety Analysis Data Handbook [44]
- Nuclear Plant Reliability Data System [45]
- Operational Performance Recording and Evaluation Data System [46]

Table 3.1 presents human error/reliability data for certain tasks [6, 41, 47, 48].

**Table 3.1** Human error/reliability data for five selected tasks

| No. | Task/error description | Errors per million operations | Performance reliability |
|---|---|---|---|
| 1 | Turning rotary selector switch to certain position | — | 0.9996 |
| 2 | Finding maintenance (scheduled) approaches in maintenance manual | — | 0.997 |
| 3 | Reading gauge incorrectly | 5,000 | — |
| 4 | Procedural error in reading instructions | 64,500 | — |
| 5 | Connecting hose incorrectly | 4,700 | — |

## 9. PROBLEMS

1. Define the following two terms:
   - Human error
   - Human reliability
2. Write an essay on the history of human error/reliability.
3. List five important facts and figures concerned with human error/reliability.
4. Discuss human error in system development stages: design, production, operation, and test.
5. Discuss the following three terms:
   - Maintenance error
   - Installation error
   - Operating error
6. Discuss the human performance effectiveness versus stress curve.
7. Discuss operator stress characteristics.
8. Discuss the four types of occupational stressors.
9. Describe the Pontecorvo method.
10. List five important sources for obtaining human error/reliability data.

## 10. REFERENCES

1. Meister, D., Human Factors in Reliability, in Reliability Handbook, edited by W.G. Ireson, McGraw-Hill Book Company, New York, 1966, pp. 400–415.
2. Hagen, E. W., editor, Human Reliability Analysis, Nuclear Safety, Vol. 17, 1976, pp. 315–326.
3. Dhillon, B. S., Singh, C., Engineering Reliability: New Techniques and Applications, John Wiley and Sons, New York, 1986.
4. Dhillon, B. S., Human Reliability: with Human Factors, Pergamon Press, Inc., New York, 1986.
5. Beecher, H. K., The First Anaesthesia Death with Some Remarks Suggested by It on the Fields of the Laboratory and the Clinic in the Appraisal of New Anaesthetic Agents, Anaesthesiology, Vol. 2, 1941, pp. 443–449.
6. Williams, H. L., Reliability Evaluation of the Human Component in Man-Machine system, Electrical Manufacturing, April 1958, pp. 78–82.
7. Shapero, A., Cooper, J. I., Rappaport, M., Shaeffer, K. H., Bates, C. J., Human Engineering Testing and Malfunction Data Collection in Weapon System Programs, WADD Technical Report No. 60–36, Wright-Patterson Air Force Base, Dayton, Ohio, February 1960.
8. Levan, W. I., Analysis of the Human Error Problem in the Field, Report No. 7–60 -932004, Bell Aero-systems Company, Buffalo, New York, June 1960.
9. Special Issue on Human Reliability, IEEE Transactions on Reliability, Vol. 22, August 1973.
10. Dhillon, B. S., On Human Reliability: Bibliography, Microelectronics and Reliability, Vol. 20, 1980, pp. 371–373.
11. Lee, K. W., Tillman, F. A., Higgins, J. J., A Literature Survey of the Human Reliability, Component in a Man-Machine System, IEEE Transactions on Reliability, Vol. 37, 1988, pp. 24–34.
12. Dhillon, B. S., Yang, N., Human Reliability: A Literature Survey and Review, Microelectronics and Reliability, Vol. 34, 1994, pp. 803–810.
13. Rook, L. W., Reduction of Human Error in Industrial Production, Report No. SCTM 93–63 (14), June 1962. Sandia National Laboratories, Albuquerque, New Mexico.
14. Majesty, M. S., Personnel Subsystem Reliability for Aerospace Systems, Proceedings of the National Aerospace Systems Reliability Symposium, April, 1962, pp. 49–55.
15. Meister, D., Human Factors: Theory and Practice, John Wiley and Sons, New York, 1971.
16. Willis, H. R., The Human Error Problem, Report No. M-62-76, Martin/Denver Company, Denver, Colorado, June 1962.
17. Kohn, L. T., Corrigan, J. M., Donaldson, M. S., editors, To Err Is Human: Building a Safer Health System, Report, Institute of Medicine, National Academy Press, Washington, D.C., 1999.
18. Robinson, J. E., Deutsch, W. E., Rogers, J. G., The Field Maintenance Interface Between Human Engineering and Maintainability Engineering, Human Factors, Vol. 12, No. 3, 1970, pp. 253–259.
19. Organizational Management and Human Factors in Quantitative Risk Assessment, Report No. 333/1992 (Report 1), British Health and Safety Executive (HSE), London, 1992.
20. Dhillon, B. S., Reliability Technology in Health Care Systems, Proceedings of the IASTED Int. Symp. Comp. Adv. Technol. Med., Health Care Bioeng., 1990, pp. 84–87.
21. Roughead, E. E., et al., Drug-Related Hospital Admissions: A Review of Australian Studies Published 1988–1996, Med. J. Aust., Vol. 168, 1998, pp. 405–408.

22. Sauer, D., Campbell, W. B., Potter, N. R., Askren, W. B., Relationships Between Human Resource Factors and Performance on Nuclear Missile Handling Tasks, Report No. AFHRL-TR-76-85/AFWL-TR-76-301, Air Force Human Resources Laboratory/Air Force Weapons Laboratory, Wright-Patterson Air Force Base, Ohio, 1976.
23. Bogner, M. S., Medical Devices: A New Frontier for Human Factors, CSERIAC Gateway, Vol. IV, No. 1, 1993, pp. 12–14.
24. Wechsler, J., Manufacturers Challenged to Reduce Medication Errors, Pharmaceutical Technology, February 2000, pp. 14–22.
25. Short, T. G., O'Regan, A., Lew, J., Oh, T. E., Critical Incident Reporting in An Anesthetic Department Quality Assurance Programme, Anesthesia, Vol. 47, 1992, pp. 3–7.
26. Christensen, J. M., Howard, J. M., Field Experience in Maintenance, In Human Detection and Diagnosis of System Failures, edited by J. Rasmussen and W. B. Rouse, Plenum Press, New York, 1981, pp. 111–133.
27. Kenny, G. C., Spahn, M. J., Amato, R. A., The Human Element in Air Traffic Control: Observations and Analysis of Performance of Controllers and Supervisors in Providing Air Traffic Control Separation Services, Report No. MTR-7655, METREK Div., MITRE Corporation, New York, December 1977.
28. Nobel, J. L., Medical Device Failures and Adverse Effects, Pediat-Emerg. Care, Vol. 7, 1991, pp. 120–123.
29. Cornell, C. E., Minimizing Human Errors, Space/Aeronautics, March 1968, pp. 71–81.
30. Willis, H. R., Human Error: Cause and Reduction, Presented to the Joint Meeting of the Midwest Human Factors Society and National Safety Council, Chicago, November 1962.
31. Rigby, L. V., Cooper, J. I., Problems and Procedures in Maintainability, Report No. ASD-TNN-61-126, Aerospace Medical Laboratory, Wright-Patterson Air Force Base, Ohio, October 1961.
32. Rigby, L. V., The Sandia Human Engineering Rate Bank (SHERB), Paper Presented at the Man-Machine Effectiveness Analysis Symposium, Los Angeles Human Factors Society, University of California at Los Angles, 1967.
33. Altman, J. W., Classification of Human Error, in Proceedings of the Symposium on Reliability of Human Performance in Work, edited by W.B. Askren, Report No. AMRL-TR-67-88, Aerospace Medical Research Laboratories, Wright-Patterson Air Force Base, Ohio, May 1967.
34. Beech, H. R., Burns, L. E., Sheffield, B. F., A Behavioral Approach to the Management of Stress, John Wiley and Sons, New York, 1982.
35. Meister, D., Human Factors in Reliability, in Reliability Handbook, edited by W.G. Ireson, McGraw Hill Book Company, New York, 1966, pp. 400–415.
36. Pontecorvo, A. B., A Method of Predicting Human Reliability, Proceedings of the 4th Annual Reliability and Maintainability Conference, 1965, pp. 337–342. This proceeding was published by Spartan Books, Washington, D.C.
37. Meister, D., Human Reliability, In Human Factors Review, edited by F.A. Muckler, Human Factors Society, Santa Monica, California, 1984, pp. 13–53.
38. Meister, D., Human Reliability Data Base and Future Systems, Proceedings of the Annual Reliability and Maintainability Symposium, 1993, pp. 276–280.
39. Dhillon, B. S., Human Error Data Banks, Microelectronics and Reliability, Vol. 30, 1990, pp. 963–971.
40. Topmillar, D. A., Eckel, J. S., Kozinsky, E. J., Human Reliability Data Bank for Nuclear Power Plant Operations: A Review of Existing Human Reliability Data Banks, Report No. NUREG/CR2744/1, U.S. Nuclear Regulatory Commission, Washington, D.C., 1982.
41. Dhillon, B. S., Design Reliability: Fundamentals and Applications, CRC Press, Inc., Boca Raton, Florida, 1999.
42. Aviation Safety Reporting Program, FAA Advisory Circular No. 00-46B, Federal Aviation Administration (FAA), Washington, D.C., June 15, 1979.
43. Munger, S. J., Smith, R. W., Payne, D., An Index of Electronic Equipment Operability: Data Store, Report No. C43-1/62 RP(1), American Institute for Research, Pittsburgh, Pennsylvania, 1962.
44. Gertman, D. I., Blackman, H. S., Human Reliability and Safety Analysis Data Handbook, John Wiley and Sons, New York, 1994.
45. Reporting Procedures Manual for the Nuclear Plant Reliability Data System (NPRDS), Southwest Research Institute, San Antonio, Texas, December 1980.
46. Urmston, R., Operational Performance Recording and Evaluation Data System (OPREDS), Descriptive Brochures, Code 3400, Navy Electronics Laboratory Center, San Diego, California, November 1971.
47. Boff, K. R., Lincoln, J. E., Engineering Data Compendium: Human Perception and Performance, Vols. 1–3, Armstrong Aerospace Medical Research Laboratory, Wright-Patterson Air Force Base, Ohio, 1988.
48. Stewart, C., The Probability of Human Error in Selected Nuclear Maintenance Tasks, Report No. EGG-SSDC-5580, Idaho National Engineering Laboratory, Idaho Falls, Idaho, 1981.

# Chapter 4

# USABILITY ENGINEERING LIFE CYCLE STAGES AND IMPORTANT ASSOCIATED AREAS

## CONTENTS



## 1. INTRODUCTION

The principles of human factors have been applied during product design and development for many decades. Continuous advances in technology (e.g., microprocessors) are allowing designers to extend the number and scope of product functions, thus resulting in more complex and sophisticated products. In turn, these are creating a greater potential for usability problems and user complaints. In order to overcome difficulties such as these, more and more engineering product manufacturers are turning to the practice of usability engineering with systematic research and design approaches, specific design methods, and extensive databases on human performance and characteristics [1].

Usability engineering is not a single-time effort where all the user interface issues are settled prior to releasing the product, but it is a set of activities that usually take place throughout the product life cycle, with a significant number of these activities being performed during early product stages prior to the design of the user interface [2]. Nonetheless, the usability engineering life cycle itself is made up of many stages and their

ISBN: 1-58883-084-5 / $35.00

understanding is absolutely essential to have user-friendly or effective products. Also, there are many usability-related areas that play an important role during the usability engineering life cycle in producing user-effective items.

This chapter discusses usability engineering life cycle stages in significant depth and presents a number of areas associated with the usability engineering life cycle.

## 2. USABILITY ENGINEERING LIFE CYCLE STAGES

Over the years, professionals working in the usability area have divided the usability engineering life cycle into many stages [1–6]. For our purpose, we have divided it into the following 11 distinct stages [2]:

- Knowing the potential users.
- Performing competitive analysis.
- Establishing usability goals.
- Carrying out parallel designs.
- Performing participatory design.
- Coordinating the total interface design.
- Applying guidelines.
- Prototyping.
- Performing interface evaluation.
- Performing iterative design.
- Obtaining data from actual field use.

Each of the aforementioned stages is discussed below, separately.

### 2.1. Knowing the Potential Users

This is the first stage of the usability engineering life cycle and is concerned with knowing the potential users of the product under consideration. Two factors that have greatest impact on usability are individual user characteristics and variability in tasks. Individual user characteristics include items such as users' work experience, age, educational level, and work environment. The variability in tasks is basically concerned with the changes in task requirements from one extreme to another. In order to determine the variations in task requirements, it is essential to perform task analysis. The performance of task analysis can lead to information such as the items listed below.

- A list of tasks users desire to accomplish with the product under consideration.
- The communication needs of users while performing the task or preparing to perform the task.
- All of the information the users will require to perform the desired tasks.
- The identification of steps involved in performing the desired tasks.
- The reports to be produced.

### 2.2. Performing Competitive Analysis

This is concerned with performing analysis of competing products with respect to their usability. Existing competing products are easy to analyze from the usability standpoint. Information on their usability-related strengths and weaknesses can be quite useful for the product under consideration. In the event of having many competing products, a comparative analysis of their user-interface issues can be performed.

All in all, competitive analysis is useful in obtaining information on issues that work best and those that should be avoided.

### 2.3. Establishing Usability Goals

The third stage of the usability engineering life cycle is concerned with developing usability-related goals to be pursued. Past experiences indicate that it is reasonably easy to develop usability goals for new versions of already existing products or for products with clearly defined competitors in the market. In such situations, the minimum acceptable usability goals usually equal the current usability levels.

On the other hand, when a completely new product is being developed without any competition, the establishment of usability goals is much more difficult.

### 2.4. Carrying Out Parallel Designs

This is concerned with carrying out the preliminary design of the same issues by several different designers simultaneously or in parallel. Subsequently, the most appropriate design is chosen, and then is subjected to more detail usability-related activities. It is important to note that, for the best results, the involved designers must work independently.

### 2.5. Performing Participatory Design

Although the first stage of the usability engineering life cycle is concerned with knowing the potential users, at that stage, the users cannot provide effective input on all issues in question. Thus, during the actual design, it is important that designers have access to a pool of representative users.

All in all, user participation during the actual design can help eliminate many usability-related problems because users raise questions that designers and their associates may not have thought of asking.

### 2.6. Coordinating the Total User Interface

It may be said that consistency is one of the most important usability characteristics. Thus, it must apply across the total user interface. Past experiences indicate that to have the total interface consistency, it is necessary to have some centralized authority coordinating various activities of the interface for each development project.

Consistency is not just measured at a single point in time, but applies over all subsequent releases of a product. All in all, it may be added that the more consistent the total user interface, the lesser the chances for user-related problems.

### 2.7. Applying Guidelines

This is concerned with applying guidelines containing well-known principles for user interface design in the development project. There are many different types of guidelines: general, specific, etc. The main difference between usability guidelines and standards is that guidelines provide advice about the usability characteristics of the interface, whereas standards specify how the interface should appear to users. A number of general user interface guidelines are given in Refs. [7–9].

### 2.8. Prototyping

Prototypes of the final systems are developed to evaluate early usability. A prototype may simply be described as a form of design specification frequently used to communicate the final design to the developer. All in all, the basic idea for prototyping is to save time and money to develop something that can be tested with actual users.

### 2.9. Performing Interface Evaluation

This is concerned with evaluating the user interface with respect to usability. Over the years, various methods for user interface testing have been developed. Their application depends on various factors including the preference of the testing individual. Although every effort should be made to use the most appropriate test method, the advantages of employing some reasonable method to evaluate a user interface rather than releasing it without an evaluation are much greater than the incremental benefits of using the exact method. The subject of usability testing is discussed in detail in Chapter 6.

### 2.10. Performing Iterative Design

A new version of the user interface design can be produced by examining the usability related problems and opportunities highlighted by the empirical testing. However, it should be noted that during the iterative design process it may not be possible to test each and every successive design version with real users. Nonetheless, iterations are a reasonably good approach to evaluate design ideas, simply by trying them out in a concrete design. Subsequently, the design can be presented for examination by usability experts, expert users, or consultants, and/or it can be subjected to heuristic analysis. All in all, past experiences indicate that in order to achieve maximum benefits, it is recommended to plan for multiple iterations.

### 2.11. Obtaining Data from Actual Field Use

After the product is put into field use, various types of real life usability-related data are collected. There are various ways and means for collecting such data. The main objective of collecting these data is to compare the actual user-interface performance with the predicted one, as well as to use these data in designing the next user interface version and future products.

## 3. BASIC FEATURES OF DESIGN FOR USABILITY AND USABILITY ACTIONS DURING SYSTEM DESIGN PHASES

This section basically considers the same thing as Section 4.2 but from a different perspective. Nonetheless, Figure 4.1 presents basic features of design for usability [10]. The user-centred design is focused, right from the start, on users and tasks. It simply means that designers must learn about all potential users and the tasks to be performed by them. This takes place prior to the start of design work, and design for usability begins by creating a usability specification document.

The participative design calls for the participation of users as members of the design team. The participation of these users is very important, particularly during the early formulation phase of the design, and when creating the usability specification. These users will need to be shown a range of possibilities and alternatives by means of mock-ups and simulations, in order for them to provide useful input.

The feature "experimental design" means that early in the development process, the potential users should conduct pilot trials using simulations and prototypes. Whenever it is feasible, the alternative versions of key features and interfaces are also simulated or prototyped for performing comparative usability-related studies. All in all, at the end, assessments are made with respect to ease of learning and usability, as well as associated difficulties.

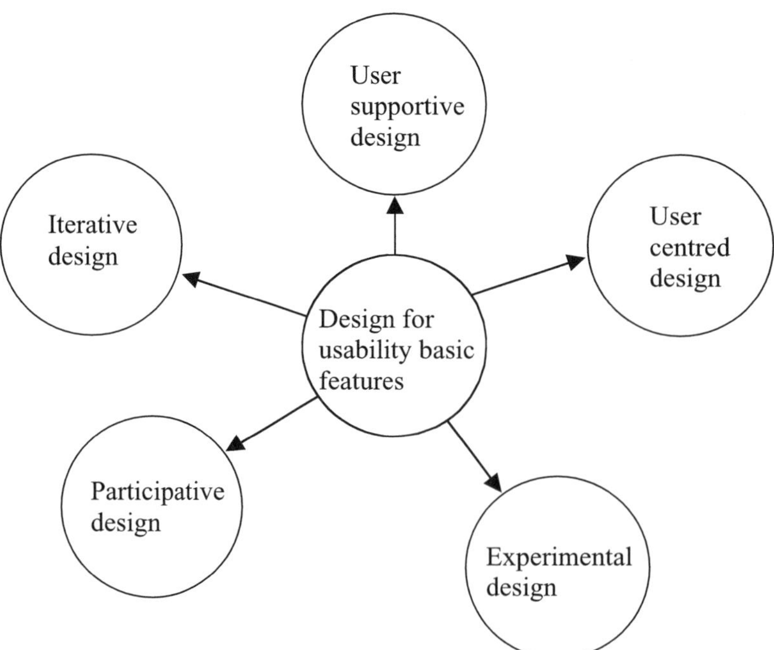


**Fig. 4.1** Fundamental features of design for usability.

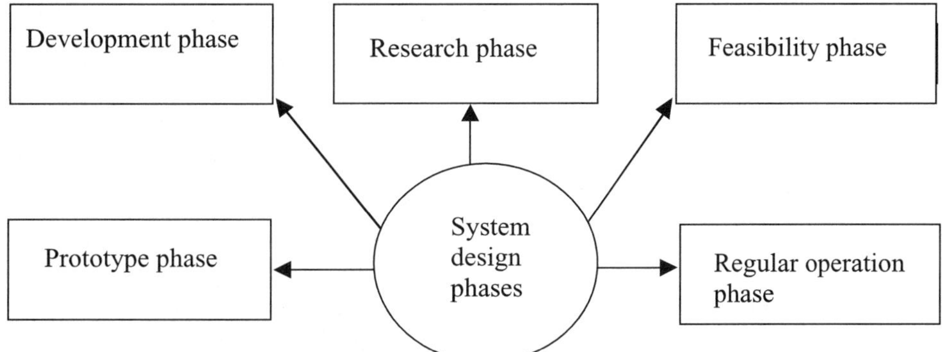


**Fig. 4.2** System design phases

The iterative design calls for carrying out test, measure, and redesign as a regular cycle until effectively satisfying the usability specification.

The user supportive design feature is usually left until a very late stage in the design process and deals with user support items such as training, manuals, and "help" systems.

System design can be divided into five distinct phases as shown in Fig. 4.2 [10]. The usability-related actions taken during each of these phases are presented below.

- **Feasibility Phase**
  Some of the usability-related actions taken during this phase are as follows:
  - Defining the range of potential users.
  - Defining the range of tasks and environments.
  - Determining if the requirements are met by proposals.
  - Establishing usability goals.
  - Creating and formalizing the usability specification.
  - Performing preliminary function and operation analyses.
- **Research Phase**
  Some of the usability-related actions that belong to this phase are as follows:
  - Performing studies of human capabilities with regard to system operational concepts.
  - Using pilot studies in the field to determine the operational requirements of users.
  - Using pilot studies in the field to determine all possible effects upon organizational and social structure.
- **Development Phase**
  Probably the greatest number of usability-related tasks are performed in this phase. Some of those tasks are as follows:
  - Performing analysis of all tasks and functions involving humans.
  - Designing all human factors aspects of equipment and workplaces.
  - Checking all design ideas against available human–related data.
  - Testing subsystem sections with samples of likely users.
  - Performing iterative design.
  - Developing user training schemes.
- **Prototype Phase**
  Some of the usability-related actions taken during this phase are as follows:
  - Conducting thorough laboratory evaluation by using samples of most likely users.
  - Conducting field trials with actual users under realistic environments.
  - Performing iterative design as appropriate.
- **Regular Operation Phase**
  Some of the usability-related tasks associated with this phase are as follows:
  - Providing for user support, i.e., training, "hot-line" for help, etc.
  - Gathering usability evaluation-related data.
  - Feeding back appropriate data to all concerned.

## 4. USABILITY PERFORMANCE MEASURES AND MYTHS

Over the years, various usability performance measures have been proposed for various purposes. Table 4.1 presents some of the most important measures [11–12]. A selection can be made from measures such as these to satisfy a specific need.

Over the years, various professionals have come across many myths regarding usability. Some of these myths are as follows [13]:

- **Myth 1: Usability lengthens development time and increases development costs.**
  In reality, the practice of usability principles can reduce overall product development time and cost by eliminating or significantly reducing the need for rework and updates. In fact, various industrial studies show that each dollar spent on user-related efforts during the product design phase saves around $10 on problem eradication during the product development phase, or $100 or more in rework after the release of product for use. Furthermore, it is believed that around 80% of all maintenance-related costs are due to overlooked user needs and the remaining 20% are due to bugs.
- **Myth 2: Usability is really just common sense.**
  In reality, the practice of usability requires an appropriate level of knowledge, skill, and experience.
- **Myth 3: Usability can be handled in help/training/documentation.**
  In reality, it is not wise to depend on help/training/documentation to overcome usability-associated problems. For example, some surveys of users with respect to the use of documentation and reference materials revealed the following results:
  - Only 9% read the printed "User manual/guide" and 80% skim it.
  - Only 11% read printed reference documentation and 46% skim it.
  - Only 7% read online materials and 54% skim them.

  All in all, it does not mean that help/training/documentation should be ignored, but it simply means that total reliance on them to overcome usability problems is not a wise course of actions.
- **Myth 4: Developers familiar with guidelines design good user interfaces.**
  In reality, guidelines can simplify the design process by reducing the look-and-feel-related decisions, but they only address a small percentage of questions raised during interface development. All in all, it may be added that the adherence to guidelines during user interface development is certainly very useful, but their limitations must also be considered with care.
- **Myth 5: Usability testing is not required if the development team has been involved with the user for a long time.**
  In reality, usability testing is always a good idea, irrespective of the length of the association, because changes in user needs occur frequently.
- **Myth 6: The user interface is just a matter of adding good graphics to make the application appealing.**
  In reality, although it is nice to have good graphics to make the application appealing, the performance of a given task efficiently and with ease is also very important.

**Table 4.1** Important usability performance measures

| No. | Measure |
|---|---|
| 1 | Number of times the user loses control of the system. |
| 2 | Time taken to accomplish task. |
| 3 | Number of times users need to work around a problem. |
| 4 | Ratio of successes to failures. |
| 5 | Number of times the interface misleads the user. |
| 6 | Percentage or number of competitors that do better than existing product. |
| 7 | Frequency of help or documentation use. |
| 8 | Number of users preferring the system. |
| 9 | Total time spent using help or documentation. |
| 10 | Number of times the user expresses frustration or satisfaction. |
| 11 | Number of good and bad features recalled by users. |
| 12 | Total time spent in errors. |

- **Myth 7: Usability is user interface design.**
  In reality, usability is much more than this, and involves a careful identification of users, their capabilities and goals, the tasks they must perform to achieve their goals, and the context in which they will be using the product.

## 5. FACTORS AFFECTING USABILITY WITHIN ORGANIZATIONS AND GUIDELINES FOR MANAGING THE POLITICS OF USABILITY

Over the years, professionals working in the area of usability have identified many organizational factors that can considerably affect usability within organizations. Most of these factors are shown in Fig. 4.3 [14].

The practice of usability is subject to politics just like any other issue. A set of useful guidelines for managing the politics of usability effectively is as follows [15]:

- **Gain support from top-level management.** This should be accomplished through means as considered appropriate by emphasizing factors such as examples of general benefits experienced by other organizations, benefits in terms of cost savings from other similar projects, and personal experiences with the practice of usability concepts.
- **Be as flexible as possible.** It means rigid usability ideas may not be that easy to sell to others. Past experiences show that flexibility with respect to various usability-related issues has proven to be a good tool to win others over.
- **Educate your users and keep them on your side.** This is basically concerned with informing users about better product usability and encouraging them to be more demanding in this regard.
- **Tactfully exploit your communication and inter-personal skills.** Build good relationships with clients and others, and then tactfully promote usability. Furthermore, ensure that your reports, presentations, and documentation adhere to good usability principles.
- **Aim for usability to be "institutionalized".** Institutionalization of usability is necessary for its success. This requires allocation of adequate resources, use of structured and consistent usability methods, etc.
- **Become a member of the usability community.** Join the appropriate professional societies to share usability-related information, experience, and problems. In turn, this may help the handling of usability issues more effectively in one's own organization.
- **Capitalize on the advantages of multi-disciplinary teams.** Past experiences indicate that usability teams work best if they are multi-disciplinary. As per many usability professionals, a team with members having knowledge in areas such as psychology, computer science, and graphic design has proven to be very effective.

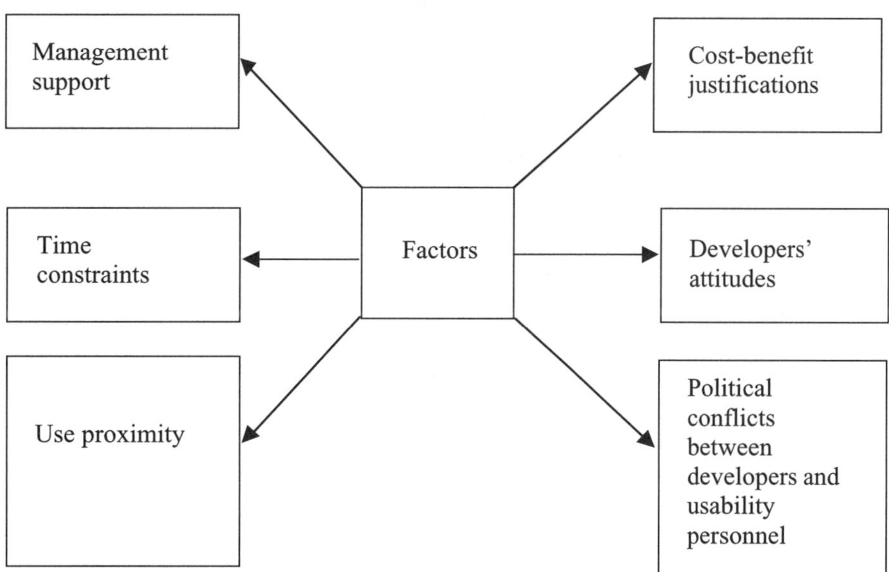


**Fig. 4.3** Important usability affecting factors within organizations

## 6. NON-FUNCTIONAL SYSTEM DEVELOPMENT PROJECT REQUIREMENTS AND THEIR IMPACT ON USABILITY AND FUNCTIONS AND QUALITIES OF A USABILITY ENGINEER

There are many non-functional requirements or factors in a system development project that may impact usability. Some of these requirements and their potential impacts on usability–listed in parentheses–are team membership (failure to include usability specialists as members of the team will compromise attention on usability), reliability (a need for highly reliable systems may require a complex structure that is reflected later in user interface effectiveness), project size (large projects are more difficult to manage and coordinate, and may lead to attention on only the most critical usability-related issues), economics (the technology cost of an effective user interface may be too high to include it in the design), portability (the requirement to build multiple compatible versions of a system may lead to the development of a "lowest common denominator", thus lower usability), and legacy systems (earlier commitment to already existing hardware/software platforms may very well over constrain space for design, thus eliminating options that would improve usability significantly) [15].

Today, the term "usability engineer" is used interchangeably with the term "usability specialist". This individual performs various functions and some of those are as follows [16]:

- Collecting data on product users.
- Performing analysis of the user needs.
- Eliminating the snags that frustrate product users.
- Interacting with engineers and others to identify issues that affect customers or users.
- Determining best choices for users.
- Predicting user needs.
- Testing hypotheses with respect to the prediction of user needs.
- Creating paper prototypes of the design.
- Making products user friendly.

Just like in the case of any other professional, the usability engineer must also possess certain qualities. Some these qualities are presented in Table 4.2.

## 7. USABILITY BENEFITS

In general, there are many benefits of good product usability. Some of these are as follows [17]:

- Reduction in training costs.
- Reduction in development costs.
- Reduction in maintenance costs.
- Improvement in productivity and efficiency.

**Table 4.2** Usability engineer's qualities

| No. | Quality |
|---|---|
| 1 | Good usability-related experience. |
| 2 | Flexibility. |
| 3 | Tact. |
| 4 | Good humor. |
| 5 | Ability to recognize different points of view. |
| 6 | Effective communication skills. |
| 7 | Self-confidence. |
| 8 | Ability to reason. |
| 9 | Good listening ability. |
| 10 | Good emotional control. |
| 11 | Good analytical ability. |
| 12 | Persuasiveness. |

- Reduction in support costs.
- Litigation deterrence.
- Improvement in competitive edge.
- Better recognition in the media.
- Reduction in documentation costs.

Important usability advantages, specifically from producers' and users' perspectives, are presented below, separately.

### Producers

- **Better quality.** It means improving usability will result in improvement in users' perception with respect to overall product quality.
- **More competitive product.** It simply means that easy-to-use products are more competitive and receptive in the market.
- **Lower cost.** It means the effective consideration of user requirements during the product development phase can reduce the need for costly late re-design work, thus lowering overall cost.
- **Reduction in support need.** Most of the times the user support needs are due to usability-related issues, and improvement in usability can reduce demands on support.

### Users

- **Greater productivity.** It means user-friendly products allow users to efficiently achieve their set goals, rather than waste hours struggling with the product user interface.
- **Greater satisfaction.** It means users will be more satisfied and positive toward a product rather than being fearful of using it.
- **Lesser need for training/support.** It means user-friendly products improve learning and can reduce the time users need for training, seeking support, or filtering through relevant documentation.

## 8. PROBLEMS

1. Discuss usability engineering life cycle stages.
2. Discuss the term "iterative design" with respect to usability.
3. What are the fundamental features of design for usability? Discuss each of these features in detail.
4. List usability actions during system design phases.
5. Discuss the term "user centred design".
6. List at least ten measures that can be used to measure usability performance.
7. Discuss at least five myths that are often associated with usability.
8. What are the advantages of usability?
9. What are the factors that affect usability within organizations?
10. List the important functions and qualities of usability engineers.

## 9. REFERENCES


1. Wiklund, M. E., How to Implement Usability Engineering, Medical Devices and Diagnostic Industry, September 1993, pp. 68–73.
2. Nielsen, J., Usability Engineering, Academic Press, Inc., Boston, Massachusetts, 1993.
3. Mayhew, D. J., The Usability Engineering Lifecycle, Morgan Kaufmann Publishers, San Francisco, 1999.
4. Shackel, B, Usability: Context, Framework, Definition, Design, and Evaluation, in Human Factors for Informatics Usability, edited by B. Shackel, Cambridge University Press, Cambridge, New York, 1991, pp. 21–37.
5. Tyldesley, D. A., Employing Usability Engineering in the Development of Office Products, The Computer Journal, Vol. 31, No. 5, 1988, pp. 431–436.
6. Gould, J. D., Lewis, C., Designing for Usability: Key Principles and What Designers Think, Communications of the ACM, Vol. 28, No. 3, 1985, pp. 300–311.
7. Smith, S. L., Mosier, J. N., Design Guidelines for Designing User Interface Software, Report No. MTR-10090, The MITRE Corporation, Bedford, Massachusetts, USA, 1986.

8. Brown, C. M. L., Human-Computer Interface Design Guidelines, Ablex Publishing Corporation, Norwood, New Jersey, USA, 1988.
9. Mayhew, D. J., Principles and Guidelines in Software User Interface Design, Prentice Hall, Inc., Englewood Cliffs, New Jerseys, USA, 1992.
10. Shackel, B., Usability: Context, Framework, Definition, Design, and Evaluation in Human Factors for Informatics Usability, edited by Shackel, B. and Richardson, S.J., Cambridge University Press, Cambridge, U.K., 1991, pp. 19–37.
11. Whiteside, J., Bennett, J., Holtzblatt, K., Usability Engineering: Our Experience and Evolution, In Handbook of Human–Computer Interaction, edited by Helander, M., North Holland Publishing Company, Amsterdam, 1987, pp. 26–40.
12. Tyldesley, D. A., Employing Usability Engineering in the Development of Office Products, The Computer Journal, Vol. 31, No. 5, 1988, pp. 431–436.
13. Chrusch, M., Seven Great Myths of Usability, Interactions, September/October issue, 2000, pp. 13–16.
14. Kaderbhai, T., Overcoming Inertia within a Large Organization: How to Overcome Resistance to Usability, in The Politics of Usability: A Practical Guide to Designing Usable Systems in Industry, edited by L. Trenner and J. Bawa, Springer-Verlag Publishers, New York, 1998, pp. 35–47.
15. Rosson, M. B., Carroll, J. M., Usability Engineering, Morgan Kaufmann Publishers, San Francisco, 2002.
16. Crosby, O., Your are a What? Usability Engineer, Occupational Outlook Quarterly, (Washington), Winter 2000/2001, pp. 1–2.

# Chapter 5

# USABILITY EVALUATION METHODS

## CONTENTS



## 1. INTRODUCTION

Usability is becoming an important factor in the success or failure of a product. This is forcing product manufacturers to seriously consider quality of use as an integral part of overall product quality. Thus, usability-related issues are becoming more integrated in the product design process as a whole. This represents a major departure from the traditional practice of assessing the product usability at the end of the design cycle [1].

The consideration of usability issues during the product design and development process often requires speedy input from usability professionals so that all necessary measures can be taken in a timely manner. Over the years, a variety of methods have been developed for addressing usability-related issues from the start of the product design process. These include methods for evaluating user characteristics, requirements, and so on.

This chapter presents a number of methods considered useful to evaluate the usability of products during the design process. Some of these methods were originally developed for application in reliability and quality areas.

## 2. TASK ANALYSIS

This method is largely used during the specification phase of product design. Task analysis may simply be defined as the study of what a user of a product is expected to do, with respect to actions and/or cognitive processes, to achieve a task objective effectively [2]. In general, it may be added that the term "task analysis" refers to a methodology that can be performed by using many specific techniques. These techniques are basically used to evaluate the interactions between the humans (users) and the products.

Task analysis helps to break down the techniques for carrying out tasks with a product under consideration into a series of steps. Consequently, the techniques can be used to predict whether the performance of tasks in

ISBN: 1-58883-084-5 / $35.00

question will be easy or difficult, as well as the degree of effort likely to be required. The result of basic task analyses provides a list of the physical steps the user must perform effectively to complete a specific task. However, complex task analyses also take into consideration the cognitive steps associated with a task.

The number of steps required to accomplish a task may be considered as an elementary measure of task complexity. It may simply be interpreted as the fewer the steps, the simpler the task.

Individuals such as potential product users, experienced product designers, and domain experts can be quite valuable informants in performing task analysis. Some of the advantages and disadvantages of task analysis are as follows [2]:

### Advantages

- Useful with respect to prescribing potential solutions to usability-related problems.
- Requires the investigator to follow a specific procedure because of the standardization of task analysis notations.
- Useful to identify the product's design that causes the inconsistencies.

### Disadvantages

- The assumption of "expert" performance with the system in question.
- Problems with the measure of task complexity (i.e., simply counting the number of steps involved in performing a task).

This method is described in detail in Refs. [3, 4].

## 3. COGNITIVE WALKTHROUGHS

These are an approach that can be used to evaluate prototype products or systems. The basic idea behind this method is to first walkthrough the operation of an interface with individuals, and then to highlight problems within that system. Usually, the method incorporates checklists for use by developers to identify possible problems with an interface. This essentially gives a framework to developers for use in checking the system from a cognitive perspective. The framework addresses issues such as [5]:

- Assumptions with respect to user knowledge.
- Actions to be taken by users at each point in an interaction.
- The linking of the interface object components to the actions to be taken by users.

For an effective application of this method, it is essential to have a clear understanding of the potential users' characteristics.

Some of the advantages and disadvantages of cognitive walkthroughs are as follows [2]:

### Advantages

- Useful to understand the user's environment.
- Relatively quick to administer and lead directly to diagnostic and prescriptive information.
- Useful to facilitate communication among design personnel, in particular, when they are divided into requirement analysts and developers.

### Disadvantages

- A high degree of reliance on the judgement of the involved investigator.
- Lack of guidance in selecting appropriate tasks for evaluation.

The method is described in detail in Refs. [2, 5–8].

## 4. KANSEI ENGINEERING

This is a product development method that takes into consideration the desirable features of products as perceived by their end users. Usually, Kansei engineering is used roughly in the middle of the design process, i.e., prior to the writing of the user interface specification and after the defining of the information architecture.

The term "Kansei" means psychological feeling or image of a product in Japanese. Thus, Kansei engineering simply refers to the translation of users' psychological feelings about a product into perceptual design components. Occasionally, Kansei engineering is also referred to as "emotional usability" or "sensory engineering". Nonetheless, Kansei engineering involves developing a database of the keywords representing the feelings of users toward products, which are utilized to develop scales. In turn, the scales are used to evaluate products. Subsequently, various factor analysis approaches are employed to highlight those specific features of product designs that correlate with the feelings of product users.

Kansei engineering has been successfully applied to the design of items such as cars, construction vehicles, houses, and costumes in various countries, including Japan and the United States. The technique is described in detail in Refs. [9–11].

## 5. PROPERTY CHECKLISTS

This method basically lists a series of product design properties as per accepted human factors "wisdom" that will ensure that a product under consideration is usable. Normally, the checklists state the high-level properties of usable design, including consistency, compatibility, and good feedback [2]. Subsequently, they (checklists) list low-level design issues relating to these properties. Some example of these issues could be the height of characters on a computer screen, on labels on products, and specifying the display and control position.

The basic idea behind the property checklist approach is to determine if the product design conforms to the listed properties. If it does not, tasks appropriate corrective measures to avoid usability-related potential problems. Some of the advantages and disadvantages of this method are as follows [2]:

### Advantages

- Useful to preserve confidentiality because it does not involve user participants.
- It can lead directly to design solutions.
- It can be applied throughout the design process and even in the evaluation of finished products.

### Disadvantages

- It is subject to the judgement of the individual compiling the checklist in the first place.
- The effectiveness of its actual application is dependent on the accuracy of judgement of the individual using it.

This method is described in detail in Ref. [12].

## 6. EXPERT APPRAISALS

This method is based upon the opinions of an expert with respect to product usability. The expert possesses extensive experience, particularly with usability-related issues, with the product under consideration. The method may cover similar issues to the ones covered by the property checklist approach, but in a greater depth. The main reason for this is that the knowledge of the expert allows him/her to focus on really important issues in a particular context. Sometimes this method uses more than one expert. Some of the advantages and disadvantages of the method are as follows [2]:

### Advantages

- No user participants are needed.
- A useful approach for providing diagnostic and prescriptive analysis.
- It can lead directly to design solutions.

### Disadvantages

- Total reliance on expert judgement.
- No direct evidence from users that the usability problems as identified by the expert will actually lead to problems.

## 7. CO-OPERATIVE EVALUATION

This method can be used during the early prototyping stages of product design or at a later stage when an existing product is to be developed further. The method evaluates interfaces based on the use of verbal protocols with users performing specified tasks under designer's observation. More specifically, the tasks are selected by the designer and then he/she observes the user's problems in operating the system under study. The method is designed to highlight usability-related problems with early product prototypes and to assist in the design refining process. The following steps are associated with the technique:

- Recruit appropriate product users.
- Prepare tasks.
- Prepare instructions for conducting actual usability evaluation.
- Ensure the readiness of the prototype to support the defined tasks and the availability of recording facilities.
- Conduct actual evaluation sessions as appropriate.
- Record all relevant data.
- Conduct post-session interviews with the users and debrief them.
- Analyse the collected data.
- Make recommendations for design improvements.

Some of the main benefits and limitations of the method are as follows:

### Benefits

- Useful to detect usability problems early in the design process.
- Useful to improve communication between designers and users.
- The method can be used by personnel with little or no training in human factors.

### Limitations

- Unsuitable for the very early phases of design.
- Unsuitable for calculating performance data.
- Large amounts of data can be quite time consuming to analyze.

## 8. CAUSE AND EFFECT DIAGRAM (CAED)

This method was developed by Professor K. Ishikawa of Tokyo University in the 1950s for use in quality control work. It can also be used to study usability-related problems. CAED is also known as Ishikawa diagram or "fishbone diagram" because of its resemblance to the skeleton of a fish as shown in Fig. 5.1. The right-hand side of the diagram (i.e., the Fish Head) denotes effect and the left-hand side shows all the possible causes which are connected to the central line known as the "Fish Spine".

The main objective of this method is to act as a first step in problem solving by generating a comprehensive list of potential causes. In turn, this can result in identification of major causes, and thus, possible remedial measures. At a minimum, the application of the CAED approach will result in a better understanding of the problem under consideration.

A cause and effect diagram can be developed by following the steps listed below.

- Establish problem statement and brainstorm to highlight all possible causes.
- Establish classifications of major causes by stratifying into natural groupings and process steps.
- Develop the diagram by connecting the causes under appropriate process steps and fill in the problem or the effect in the diagram box (i.e., the Fish Head) on the extreme right-hand side.
- Refine cause classifications by asking questions such as:
  - What is the reason for the existence of this condition?
  - What causes this?

There are many advantages of the cause and effect diagram. Some of those are listed below.

- Useful to identify root causes.
- Useful to generate ideas.

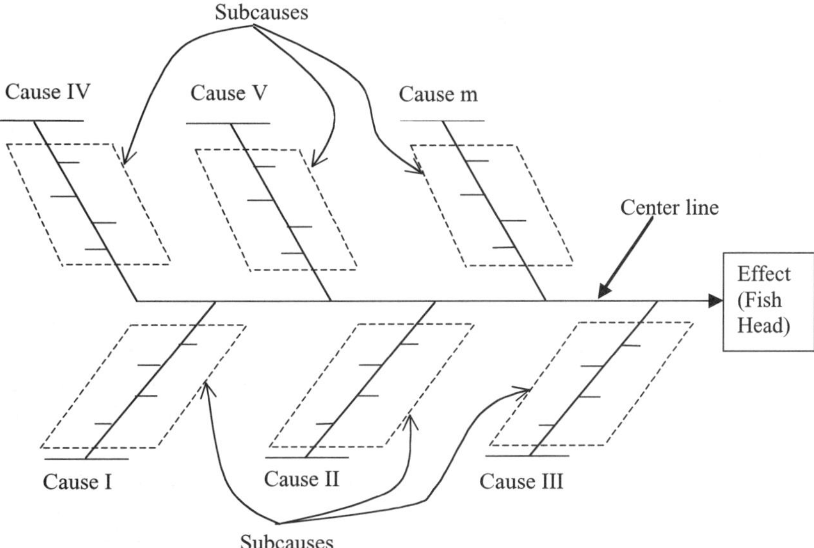


**Fig. 5.1** A cause and effect diagram for m number of causes

- A useful tool in guiding further inquiry.
- An effective tool to present an orderly arrangement of theories.

Additional information on this method may be found in Refs. [13–15].

---

## Example 5.1

Assume that a product is being designed by a company for use in an engineering system. Careful investigation revealed that it may have usability problems due to the following six main causes:

- **Cause I:** Poor design.
- **Cause II:** Poor considerations to human factors during design.
- **Cause III:** Poorly written operating instructions.
- **Cause IV:** Poor user training.
- **Cause V:** Inadequate design and development time.
- **Cause VI:** Poor user testing.

Construct a cause and effect diagram.
The cause and effect diagram for the example is shown in Fig. 5.2.

---

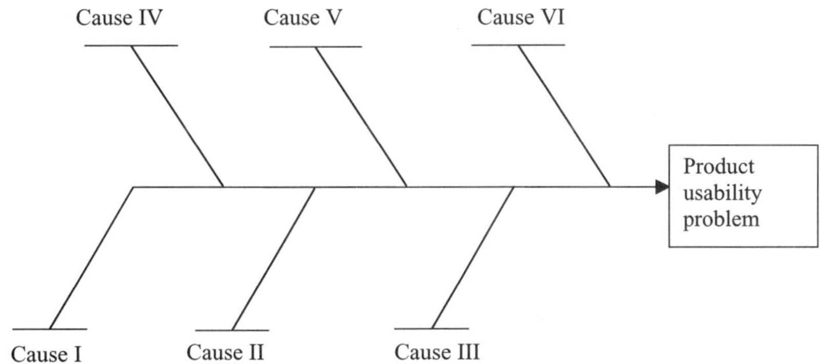


**Fig. 5.2** A cause and effect diagram for Example 5.1

## 9. PROBABILITY TREE ANALYSIS

This can be a good method for conducting usability-related task analysis by diagrammatically representing human actions and other related events in question. Diagrammatic task analysis is represented by the branches of the probability tree. More specifically, the branching limbs of the tree represent the outcome of each event (i.e., success or failure) and each branch is assigned probability of occurrence [16].

Some of the advantages of the probability tree analysis approach are flexibility to incorporate (i.e., with some modifications) factors such as emotional stress, interaction stress, interaction effects, a visibility tool, and simplified mathematical computations. The method is demonstrated by solving the following example:

---

**Example 5.2**

Assume that an individual has to perform three independent and distinct tasks $x$, $y$, and $z$ to operate or use an engineering system. Task $x$ is performed before task $y$, and task $y$ before task $z$. Furthermore, each of these tasks can either be performed correctly or incorrectly.

Develop a probability tree and obtain an expression for probability of not successfully accomplishing the overall mission (i.e., not operating the system correctly).

In this example, the individual first performs task $x$ correctly or incorrectly and then proceeds to perform task $y$. After performing task $y$ correctly or incorrectly, the same individual proceeds to perform task $z$. This task can also be performed correctly or incorrectly. This entire scenario is depicted by a probability tree shown in Fig. 5.3.

The symbols used in the figure are defined below.

$x$ denotes the event that task $x$ is performed correctly.
$\overline{x}$ denotes the event that task $x$ is performed incorrectly.
$y$ denotes the event that task $y$ is performed correctly.
$\overline{y}$ denotes the event that task $y$ is performed incorrectly.
$z$ denotes the event that task $z$ is performed correctly.
$\overline{x}$ denotes the event that task $z$ is performed incorrectly.

---

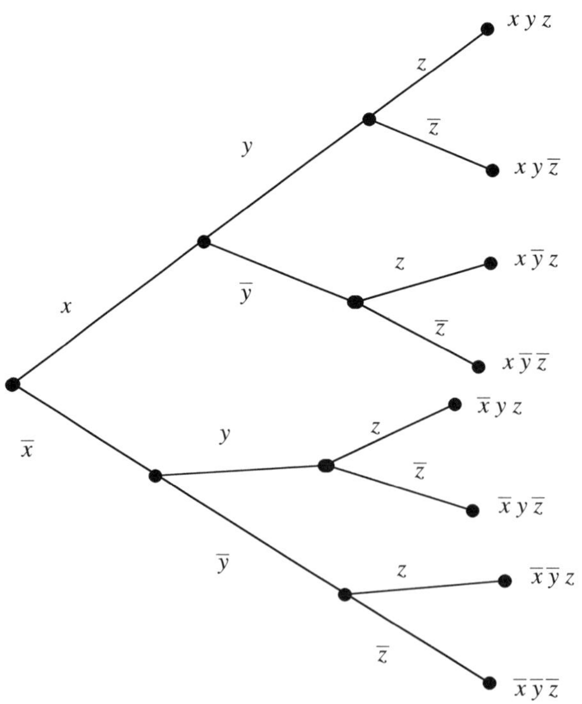


**Fig. 5.3** A probability tree for performing tasks $x$, $y$, and $z$

In Fig. 5.3, the term $xyz$ denotes operating the system successfully (i.e., overall mission success). Thus, the probability of occurrence of events $xyz$ is [17].

$$P(xyz) = P_x P_y P_z \tag{5.1}$$

where
$P_x$ is the probability of performing task $x$ correctly.
$P_y$ is the probability of performing task $y$ correctly.
$P_z$ is the probability of performing task $z$ correctly.

Similarly, in Fig. 5.3, the terms $xy\bar{z}$, $x\bar{y}z$, $x\bar{y}\bar{z}$, $\bar{x}yz$, $\bar{x}y\bar{z}$, $\bar{x}\bar{y}z$, and $\bar{x}\bar{y}\bar{z}$ denote seven distinct possibilities of not operating the system correctly or successfully. Thus, the probability of not successfully accomplishing the overall mission is

$$\begin{aligned} P_f &= P(xy\bar{z} + x\bar{y}z + x\bar{y}\bar{z} + \bar{x}yz + \bar{x}y\bar{z} + \bar{x}\bar{y}z + \bar{x}\bar{y}\bar{z}) \\ &= P_x P_y P_{\bar{z}} + P_x P_{\bar{y}} P_z + P_x P_{\bar{y}} P_{\bar{z}} + P_{\bar{x}} P_y P_z \\ &\quad + P_{\bar{x}} P_y P_{\bar{z}} + P_{\bar{x}} P_{\bar{y}} P_z + P_{\bar{x}} P_{\bar{y}} P_{\bar{z}} \end{aligned} \tag{5.2}$$

where
$P_{\bar{x}}$ is the probability of performing task $x$ incorrectly.
$P_{\bar{y}}$ is the probability of performing task $y$ incorrectly.
$P_{\bar{z}}$ is the probability of performing task $z$ incorrectly.
$P_f$ is the probability of not successfully accomplishing the overall mission (i.e., mission failure).

---

**Example 5.3**

Assume that in Example 5.2, the probabilities of the individual not performing tasks $x$, $y$, and $z$ correctly are 0.1, 0.2, and 0.05, respectively. Calculate the probability of correctly operating or using the system by the individual.

Thus, we have $P_{\bar{x}} = 0.1$, $P_{\bar{y}} = 0.2$, $P_{\bar{z}} = 0.05$.

Since $P_{\bar{x}} + P_x = 1$, $P_{\bar{y}} + P_y = 1$, and $P_{\bar{z}} + P_z = 1$, we have

$$P_x = 1 - P_{\bar{x}} \tag{5.3}$$

$$P_y = 1 - P_{\bar{y}} \tag{5.4}$$

$$P_z = 1 - P_{\bar{z}} \tag{5.5}$$

Using Equations (5.3)–(5.5) and the given data values in Equation (5.1), we get

$$\begin{aligned} P(xyz) &= (1 - P_{\bar{x}})(1 - P_{\bar{y}})(1 - P_{\bar{z}}) \\ &= (1 - 0.1)(1 - 0.2)(1 - 0.05) \\ &= 0.684 \end{aligned}$$

Thus, probability of correctly operating or using the system by the individual is 0.684.

---

## 10. FAILURE MODES AND EFFECT ANALYSIS (FMEA)

This is a very versatile method used widely in the industrial sector to analyze systems from reliability and safety aspects during the design phase. After some minor changes, the method can also be used to perform usability analysis of engineering products during the design phase.

FMEA may simply be described as a very effective approach for performing analysis of each and every potential failure mode in the system to determine effects of such modes on the total system [18]. The method was developed in the early 1950s by the United States Navy's Bureau of Aeronautics [19]. Subsequently, the National Aeronautics and Space Administration (NASA) extended it to classify each potential failure effect according to its severity and renamed it: failure mode effects and criticality analysis (FMECA) [19].

Major steps involved in performing FMEA are as follows [20]:

- Establish system boundaries and detailed requirements.
- List all system subsystems and components.
- Identify and describe each component and list all its possible failure modes.
- Assign occurrence probability/failure rate to each failure mode.
- List effect(s) of each failure mode on subsystem/system/plant.
- Enter appropriate remarks for each identified failure mode.
- Review all critical failure modes carefully and take appropriate action.

Some of the advantages of performing FMEA are reduction in cost and development time, better customer satisfaction, reduction in engineering changes, better communication among design interface personnel, identification of safety concerns to be focused on, a systematic approach to classify failures, and useful to compare alternative designs [21].

This method is described in detail in Ref. [22].

## 11. FAULT TREE ANALYSIS (FTA)

This method was developed in the early 1960s at the Bell Telephone Laboratories to conduct safety analysis of the Minuteman Launch Control System. Today, it is widely used in industry to evaluate reliability and safety of engineering systems during the design and development phase. FTA can also be used to perform usability analysis of engineering products during the design phase.

A fault tree may simply be described as a logical representation of the relationship of basic events that may cause the occurrence of a specified undesirable event, called the "Top Event", and is depicted using a tree structure with gates such as AND and OR. Four commonly used symbols in the construction of fault trees are shown in Fig. 5.4 [20, 23].

Each of these symbols is described below.

- **AND gate.** This denotes that an output fault event occurs only if all the input fault events occur.
- **OR gate.** This denotes that an output fault event occurs if any one or more input fault events occur.
- **Rectangle.** This denotes a fault event that results from the logical combination of fault events through the input of a logic gate such as AND and OR.
- **Circle.** This denotes a basic fault event or the failure of an elementary component. The values of the basic event's parameters such as occurrence probability and occurrence rate are usually obtained from empirical data.

Usually, FTA is performed by following the steps listed below [14, 20].

- Define the system and all its associated assumptions.
- Identify the undesirable event (i.e., top fault event) to be investigated (e.g., product usability problem).

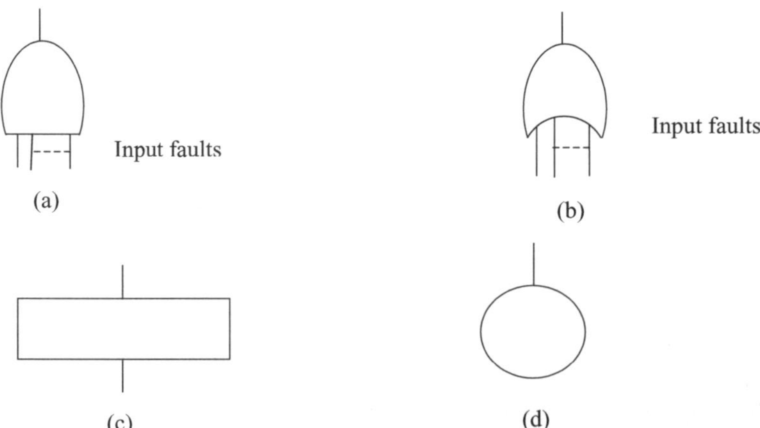


**Fig. 5.4** Basic fault tree symbols: (a) AND gate, (b) OR gate, (c) rectangle, (d) circle

- Determine all the possible causes that can result in the occurrence of the top event by using fault tree symbols such as those given in Fig. 5.4 and the logic tree format.
- Develop the fault tree to the lowest level of detail as per the needs.
- Conduct analysis of the developed fault tree with respect to factors such as understanding the logic and interrelationships among the fault paths and gaining insight into the unique modes of item faults.
- Determine the most effective corrective measures.
- Document analysis and follow-up on highlighted corrective measures.

---

**Example 5.4**

After a careful investigation, it was concluded that a product under design could have usability problems due to known basic factors: poor design and poorly written use procedures. In turn, the causes for the poor design are poor consideration to human factors, poor design management, too short development time, and poorly trained design professionals. Similarly, the causes for the poorly written use procedures are poorly trained documentation personnel, poor management, and carelessness.

Develop a fault tree for the top event "product usability problem" by using Fig. 5.4 symbols.

A fault tree for the example is shown in Fig. 5.5. The single capital alphabet letters in the figure denote corresponding fault events (e.g., A: poor consideration to human factors, and B: poorly trained design professionals).

---

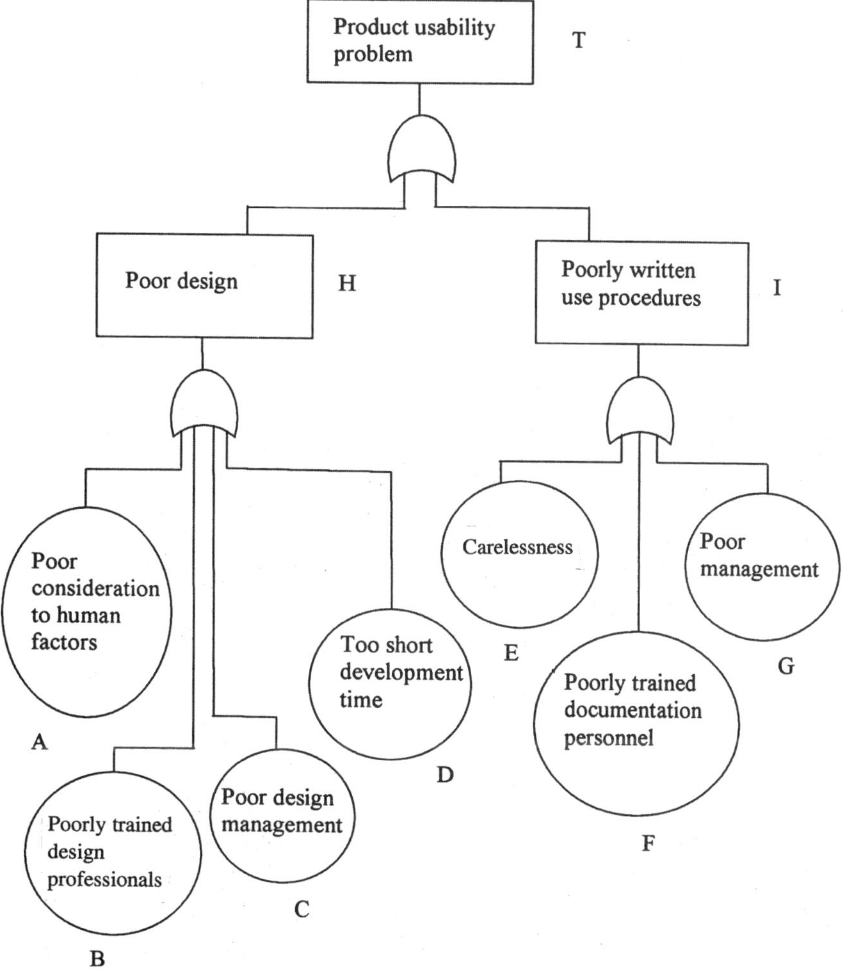


**Fig. 5.5** A fault tree for Example 5.4

### 11.1. Fault Tree Probability Evaluation

The probability of occurrence of the top event can be calculated if the occurrence probabilities of basic fault events are known. This can only be calculated by first determining the occurrence probability of the output fault events of lower and intermediate logic gates such as AND and OR. The probability of occurrence of the AND gate output fault event can be estimated by using the following equation [22]:

$$P(X_0) = \prod_{i=1}^{k} P(X_i) \tag{5.6}$$

where

- $k$ is the number of gate input fault events.
- $P(X_i)$ is the probability of occurrence of AND gate input fault event $X_i$; for $i = 1, 2, 3, \ldots, k$.
- $P(X_0)$ is the occurrence probability of the AND gate output fault event $X_0$.

Similarly, the probability of occurrence of the OR gate output fault event, $Y_0$, can be estimated by using the following equation [22]:

$$P(Y_0) = 1 - \prod_{i=1}^{k} \{1 - P(Y_i)\} \tag{5.7}$$

where

- $P(Y_0)$ is the probability of occurrence of the OR gate output fault event, $Y_0$.
- $k$ is the number of gate input fault events.
- $P(Y_i)$ is the probability of occurrence of OR gate input fault event $Y_i$; for $i = 1, 2, 3, \ldots, k$.

---

**Example 5.5**

Assume that in Fig. 5.5, the occurrence probabilities of events A, B, C, D, E, F, and G are 0.07, 0.06, 0.05, 0.04, 0.03, 0.02, and 0.01, respectively. Calculate the probability of occurrence of the top event T.

Using the specified occurrence probability values for events A, B, C, and D in Equation (5.7), the probability of occurrence of event H: poor design is

$$\begin{aligned} P(H) &= 1 - (1 - 0.07)(1 - 0.06)(1 - 0.05)(1 - 0.04) \\ &= 0.2027 \end{aligned}$$

Similarly, by substituting the given probability values for events E, F, and G into Equation (5.7), we get the following value for probability of occurrence of event I: poorly written use procedures:

$$\begin{aligned} P(I) &= 1 - (1 - 0.03)(1 - 0.02)(1 - 0.01) \\ &= 0.0589 \end{aligned}$$

By inserting the above two calculated values into Equation (5.7), we get

$$\begin{aligned} P(T) &= 1 - (1 - 0.2027)(1 - 0.0589) \\ &= 0.2496 \end{aligned}$$

Thus, the probability of the top event T (i.e., product usability problem) occurrence is 0.2496. Fig. 5.6 shows Fig. 5.5 fault tree with the above calculated and given fault event occurrence probability values.

---

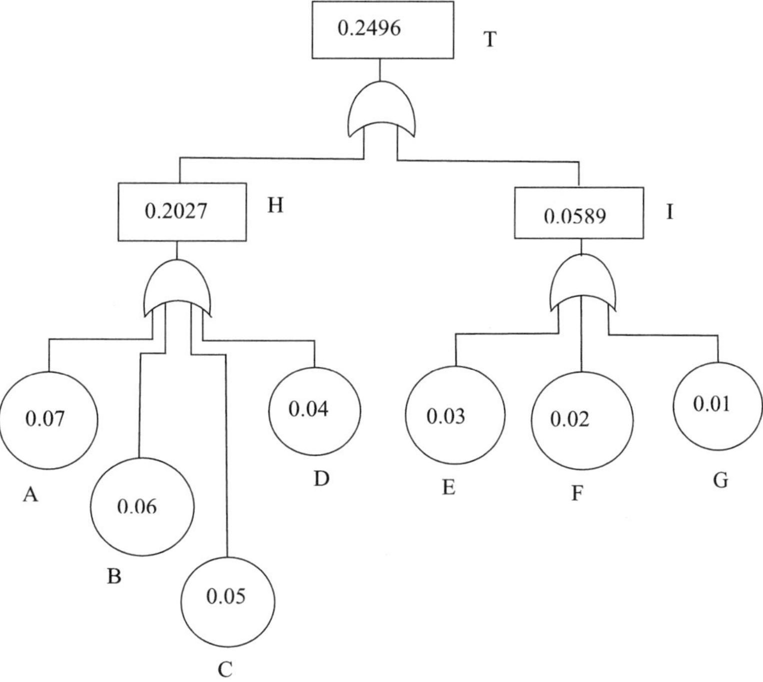


**Fig. 5.6** A fault tree with fault event probability of occurrence values

## 12. MARKOV METHOD

This is a widely used method in reliability and safety studies and it can also be used to perform various types of usability analysis during the design phase. The method is named after a Russian mathematician, Andrei Andreyevich Markov (1856–1922), and is subject to the following assumptions:

- The occurrence probability of a transition from one system state to another in the finite time interval $\Delta t$ is given by $\lambda \Delta t$, where the $\lambda$ is the transition rate from one system state to another.
- All occurrences are independent of each other.
- The transitional probability of more than one occurrence in time interval $\Delta t$ from one state to another is negligible (e.g., $(\lambda \Delta t)(\lambda \Delta t) \rightarrow 0$).

The application of this method to a usability problem is demonstrated through the following example:

---

**Example 5.6**

Assume that the usability-related difficulties associated with a system under design may be classified into two distinct categories: user error and user problem. The user error and problem rates are denoted by $\lambda$ and $\theta$, respectively. The system state space diagram is shown in Fig. 5.7. The numerals in the box, circle, and diamond denote system states. Develop expressions for system reliability, probability of user error, probability of user problem, and mean time to user error or problem by using the Markov method.

---

Using the Markov method, we write down the following equations for Fig. 5.7 [22]:

$$P_0(t + \Delta t) = P_0(t)(1 - (\lambda + \theta)\Delta t) \tag{5.8}$$

$$P_1(t + \Delta t) = (\lambda \Delta t)P_0(t) + P_1(t) \tag{5.9}$$

$$P_2(t + \Delta t) = (\theta \Delta t)P_0(t) + P_2(t) \tag{5.10}$$

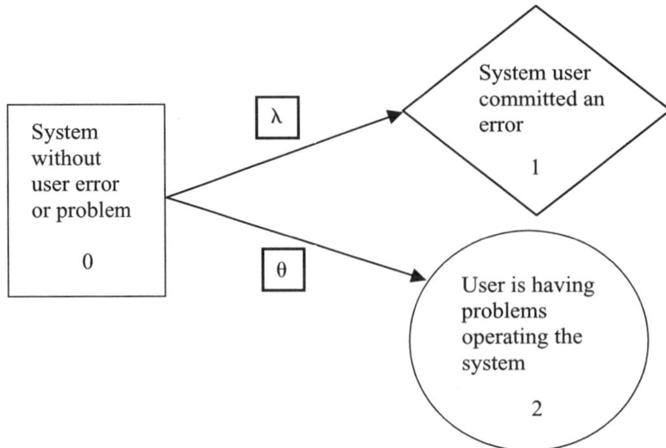


**Fig. 5.7** System state space diagram

where

| | |
|---|---|
| $t$ | is time. |
| $P_0\ (t + \Delta t)$ | is the probability of the system without user error or problem at time $(t + \Delta t)$. |
| $P_1\ (t + \Delta t)$ | is the probability that system user committed an error at time $(t + \Delta t)$. |
| $P_2\ (t + \Delta t)$ | is the probability that user is having problems operating the system at time $(t + \Delta t)$. |
| $P_i\ (t)$ | is the probability that the system is in state $i$ at time $t$; for $i = 0$ (system without user error or problem), $i = 1$ (system user committed an error), and $i = 2$ (user is having problems operating the system). |
| $\lambda\Delta t$ | is the probability of system user committing an error in finite time interval $\Delta t$. |
| $\theta\Delta t$ | is the probability of that the user is having problems operating the system in finite time interval $\Delta t$. |

From Equations (5.8)–(5.10), we get the following Equations (5.11)–(5.13), respectively.

$$\frac{dP_0(t)}{dt} + (\lambda + \theta)P_0(t) = 0 \qquad \textbf{(5.11)}$$

$$\frac{dP_1(t)}{dt} - \lambda P_0(t) = 0 \qquad \textbf{(5.12)}$$

$$\frac{dP_2(t)}{dt} - \theta P_0(t) = 0 \qquad \textbf{(5.13)}$$

At time $t = 0$, $P_0(0) = 1$ and $P_1(0) = P_2(0) = 0$.
By solving Equations (5.11)–(5.13) with Laplace transforms, we get:

$$P_0(t) = e^{-(\lambda + \theta)t} \qquad \textbf{(5.14)}$$

$$P_1(t) = \frac{\lambda}{\lambda + \theta}(1 - e^{-(\lambda + \theta)t}) \qquad \textbf{(5.15)}$$

$$P_2(t) = \frac{\theta}{\lambda + \theta}(1 - e^{-(\lambda + \theta)t}) \qquad \textbf{(5.16)}$$

Mean time to system having user error or problem is given by:

$$MTTSHUEP = \int_0^\infty P_0(t)dt = \int_0^\infty e^{-(\lambda+\theta)t}\,dt = \frac{1}{\lambda+\theta} \tag{5.17}$$

where

MTTSHUEP is the mean time to system having user error or problem.

---

**Example 5.7**

Assume that a user operating an engineering system makes 0.0004 errors/hour and has 0.0009 hourly usability-related problems. Calculate the mean time to system having user error or problem.

Substituting the given data values into Equation (5.17) yields:

$$MTTSHUEP = \frac{1}{0.0004 + 0.0009} = 769.23 \text{ hours}$$

Thus, the mean time to system having user error or problem is 769.23 hours.

---

## 13. PROBLEMS

1. Describe the following two methods:
   - Task analysis
   - Co-operative evaluation
2. What are the benefits and drawbacks of cognitive walkthroughs?
3. Describe Kansei engineering.
4. Discuss the following two usability evaluation methods.
   - Property checklists
   - Expert appraisals
5. Describe cause and effect diagram.
6. Compare the advantages and disadvantages of the following two usability evaluation methods:
   - Task analysis
   - Property checklists
7. What is the difference between failure modes and effect analysis and failure mode effects and criticality analysis?
8. Discuss four commonly used symbols in fault tree analysis.
9. What is probability tree analysis?
10. Compare probability tree analysis with fault tree analysis.

## 14. REFERENCES


1. Jordan, P. W., Thomas, B., McClelland, I. L., Issues for Usability Evaluation in Industry: Seminar Discussions, in Usability Evaluation in Industry, edited by P. W. Jordan, B. Thomas, B. A., Weerdmeester, and I. L. McClelland, Taylor and Francis, Ltd., London, 1996, pp. 237–243.
2. Jordan, P. W., An Introduction to Usability, Taylor and Francis Ltd., London, 1998.
3. Kirwan, B., Ainsworth, L. K., editors, A Guide to Task Analysis, Taylor and Francis, Ltd., London, 1992.
4. Drury, C. G., Task Analysis Methods in Industry, Applied Ergonomics, Vol. 14, No. 1, 1983, pp. 19–28.
5. Wharton, C., Bradford, J., Jeffries, R., Franzke, M., Applying Cognitive Walkthroughs to More Complex User Interfaces: Experiences, Issues, and Recommendations, Proceedings of the ACM Conference on Human Factors in Computing Systems, 1992, pp. 381–388.

6. Karat, C. M., Campbell, R., Fiegel, T., Comparison of Empirical Testing and Walkthrough Methods in User Interface Evaluation, Proceedings of the ACM Conference on Human Factors in Computing Systems, 1992, pp. 397–404.
7. Wharton, C., Rieman, J., Lewis, C., Polson, P., The Cognitive Walkthrough: A Practioner's Guide, in Usability Inspection Methods, edited by J. Nielsen and R. L. Mack, John Wiley and Sons, New York, 1994, pp. 80–100.
8. Rowley, D. E., Rhoades, D. G., The Cognitive Jog through: A Fast-paced User Interface Evaluation Procedure, Proceedings of the ACM Conference on Human Factors in Computing Systems, 1992, pp. 389–395.
9. Nagamachi, M., Kansei Engineering, Kaibundo Publishers, Tokyo, 1989.
10. Kashiwagi, K., Matsubara, A., and Nagamachi, M., A Feature Detection Mechanism of Design in Kansei Engineering, Human Interface, Vol. 9, No. 1, 1994, pp. 9–16.
11. Miyazaki, K., Matsubara, Y., Nagamachi, M., A Modeling of Design Recognition in Kansei Engineering, Japanese Journal of Ergonomics, Vol. 29, 1993, pp. 196–197.
12. Ravden, S. J., Johnson, G. I., Evaluating Usability of Human-Interfaces: A Practical Method, Ellis Horwood Publishers, Chichester, U.K., 1989.
13. Ishikawa, K., Guide to Quality Control, Asian Productivity Organization, Tokyo, 1982.
14. Mears, P., Quality Improvement Tools and Techniques, McGraw Hill Book Company, New York, 1995.
15. Kanji, G. K., Asher, M., 100 Methods for Total Quality Management, Sage Publications Ltd., London, 1996.
16. Swain, A. D., A Method for Performing a Human Factors Reliability Analysis, Report No. SCR-685, Sandia Corporation, Albuquerque, New Mexico, USA, August 1963.
17. Dhillon, B. S., Human Reliability: with Human Factors, Pergamon Press, Inc., New York, 1986.
18. Omdahl, T. P., Editor, Reliability, Availability, and Maintainability (RAM) Dictionary, American Society for Quality Control (ASQC) Press, Milwaukee, Wisconsin, 1988.
19. Jordan, W. E., Failure Modes, Effects, and Criticality Analyses, Proceedings of the Annual Reliability and Maintainability Symposium, 1972, pp. 30–37.
20. Dhillon, B. S., Singh, C., Engineering Reliability: New Techniques and Applications, John Wiley and Sons, Inc., New York, 1981.
21. Palady, P., Failure Modes and Effects Analysis, PT Publications, West Palm Beach, Florida, 1995.
22. Dhillon, B. S., Design Reliability: Fundamentals and Applications, CRC Press, Inc., Boca Ration, Florida, 1999.
23. Schroder, R. J., Fault Tree for Reliability Analysis, Proceedings of the Annual Symposium on Reliability, 1970, pp. 170–174.

# Chapter 6

## USABILITY TESTING

### CONTENTS



## 1. INTRODUCTION

Usability testing is an important research tool with its roots going back to classical experimental methodology. The range of tests can vary considerably, from true classical experiments with fairly large sample sizes and complex and sophisticated test designs to informal qualitative-based studies having only one participant [1].

Usability testing incorporates various approaches for having potential users try out a product. More specifically, in a typical usability test, the users carry out a variety of tasks with a prototype under consideration while observers monitor their actions. Usually, tests are carried out with a single user working alone or two users working jointly. Testing may involve collecting information on factors such as the paths users take to perform tasks, errors made by users, user frustrations and confusions, the degree of efficiency in performing tasks, and the degree of user satisfaction with the experience. Thus, usability testing may simply be defined as a process that uses participants representing the targeted population to determine the degree to which a product satisfies specific criteria.

This chapter presents various aspects of usability testing.

## 2. USABILITY TESTING GOALS, BENEFITS, AND LIMITATIONS

With respect to usability, the main intent is to ensure the development of products that satisfy factors such as easy to learn and use, provide satisfaction to users, and provide utility and functionality that are highly valued by the targeted population of users [2]. Thus, the overall goal of usability testing is to identify and rectify usability-related shortcomings in products and their associated support materials prior to release. More specifically, important goals of usability testing are shown in Fig. 6.1 [1]. The goal "minimize cost" is concerned with reducing the cost of service and hotline calls.

ISBN: 1-58883-084-5 / $35.00

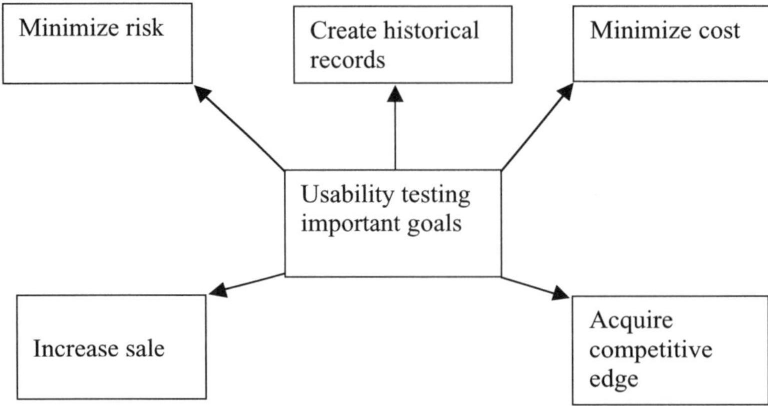


**Fig. 6.1** Important goals of usability testing

The goal "create historical records" is concerned with developing a historical record of usability benchmarks for use in future products. The goal "increase sale" is concerned with increasing sales by making the users happy with the product, who will then buy future releases of the product rather than procuring the competitors' product.

The goal "minimize risk" is concerned with reducing the risk of product failure by eradicating serious usability problems. The goal "acquire competitive edge" is concerned with gaining a competitive edge by convincing potential customers that the product under consideration is more user friendly or easier to use.

Over the years, professionals involved with usability have specifically stated many benefits and limitations of usability testing. Some of these are as follows [1, 3–4]:

### Benefits

- Overall improvement in product quality.
- Reduction in the learning curve and the training efforts.
- Wider acceptability of the product by potential customers.
- Reduction in the need for customer support services.
- Accumulation of useful usability data for future product releases.

### Limitations

- No 100% guarantee for product usability success.
- User test participants are rarely fully representative of the targeted user population.
- Impossible to simulate exact real life product use environments.

## 3. USABILITY TESTING ELEMENTS, TEST PLAN, AND TEST BUDGET

Usability testing is composed of many elements. The basic ones are as follows [1]:

- The representation of the real use environment.
- Data collection, including quantitative and qualitative performance and preference measures.
- Establishment of test objectives.
- Recommending improvements to product design.
- Using a sample of representative users who may or may not be randomly selected.
- Observing representative product users who either actually use or review a representation of the product under consideration.

A usability test plan must be properly documented prior to the start of testing. It should address issues such as test goals, date and place of the test, the number of test users required, the identification of test experimenters, the length of each test session, the degree to which test experimenters should help test users, the user aids to be made available to test users, the level of computer support required for the test, the identification of tasks to be performed by test users, the identification of test users and the approach to get hold of them, the identification of data to be collected and the approach to analyze these data, the identification of criterion for pronouncing the interface a success, the state of the product at the start of the test, and criteria for determining when users have completed each of the prescribed tasks correctly [5].

Usually, a usability test plan includes a budget for the test. Some of the major cost components of a user test budget are as follows [5, 6]:

- **Test equipment cost.** This is the estimated cost of equipment to be used during testing and analysis.
- **Usability specialist cost.** This is the estimated cost of specialists to be used to plan, run, and analyze the usability test.
- **Support personnel cost.** This is the estimated cost of administrative assistants to be used to schedule test users, enter data, and so on.
- **Facility cost.** This includes costs other than that of the test equipment cost. More specifically, the estimated cost of the usability laboratory or other rooms to be used in performing the test.
- **Test users' time cost.** This is the estimated cost of time to be spent by representative users in testing the product.

## 4. TYPES OF USABILITY TESTS

Usability tests may be divided into four distinct types as shown in Fig. 6.2 [1].

The exploratory test is performed quite early in the product development cycle, i.e., preliminary defining and designing stages of the product. The main goal of this test is to determine the effectiveness of preliminary design concepts with respect to usability. Nonetheless, the exploratory test helps to answer questions on areas such as user thinking about the product, prerequisite information required by users, user interface intuitiveness with respect to operations and navigation, user functions requiring written documentation, and the value of the product's basic functionality to users.

The assessment test is performed either early or during the middle of the product development cycle, usually after the establishment of the fundamental or high-level design or organization of the product under consideration. The main objective of this test is to complement the findings of the previous test (i.e., exploratory) by determining the usability effectiveness of lower-level operations and aspects of the product. All in all, unlike the exploratory test, assessment test users always carry out tasks instead of simply walking through and commenting upon items in question. Then, data on quantitative measures are collected, and the interactions between monitors and representative user-participants are decreased substantially because of greater emphasis on actual user behaviours.

The validation test is also known as the verification test and is often performed late in the product development cycle. More specifically, it takes place close to the release of the product. The main purpose of conducting this test is to evaluate how the product under consideration measures to some predetermined

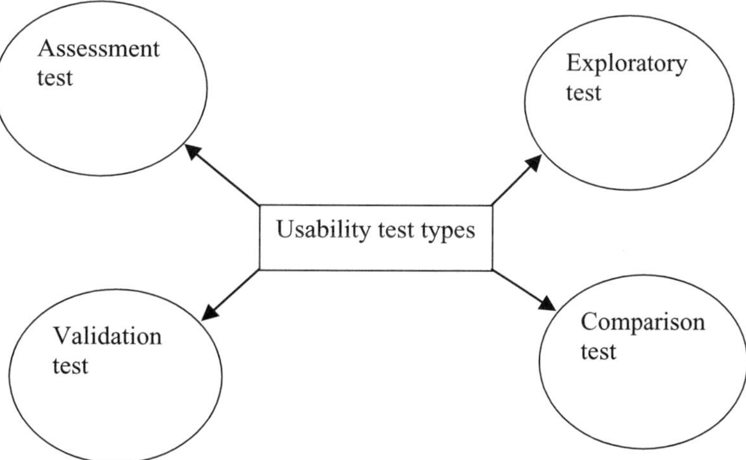


**Fig. 6.2** Types of usability tests

usability standard or benchmark. Therefore, the intent is to establish that the product satisfies such a standard or benchmark prior to its release. However, if it does not, then all possible reasons for this scenario need to be established.

It may be said that with the exception of three major factors, the validation test is basically performed in similar manner to the assessment test. These factors are existence of standards or benchmarks for the test tasks prior to the initiation of the test, very little or no interaction between user-participants and test monitors, and focus on quantitative data collection.

Finally, the comparison test is not associated with any particular point in the development cycle of a product, but it can be used in conjunction with any of the three previous tests to compare two or more alternative designs with respect to usability. Usually, for each alternative design, performance and preference data are collected and a comparison of end results is made.

## 5. USABILITY TEST PERFORMANCE STAGES

The performance of a usability test may be divided into a number of stages. For our purpose, we have divided it into six stages as shown in Fig. 6.3 [1].

The six categories include test plan development, user-participant selection, test material preparation, actual test performance, participant debriefing, and data analysis and recommendations. Each of these stages is described in detail below, separately [1, 7–13].

### 5.1. Test Plan Development

Although the subject of test plan was discussed earlier in the chapter, for the sake of completion, it is discussed again from a different angle. Nonetheless, it may be said that a test plan is the foundation for the entire test. Thus, it must be developed with great care and in a comprehensive manner.

Some of the important reasons for developing a comprehensive test plan are to serve as the blueprint for the test, to provide a focal point for the test, to serve as the main communication vehicle among the main developer, the test monitor, and the rest of the development team, and to describe or imply required resources, both internal and external.

The format of a test plan may vary from one usability test to another and from one organization to another. However, a typical test plan usually contains sections on topics such as purpose (general), problem statement/test objective, user profile, approach (test design), task list, test environment/equipment, test monitor role, data collection (evaluation measures), and report contents and presentation [1].

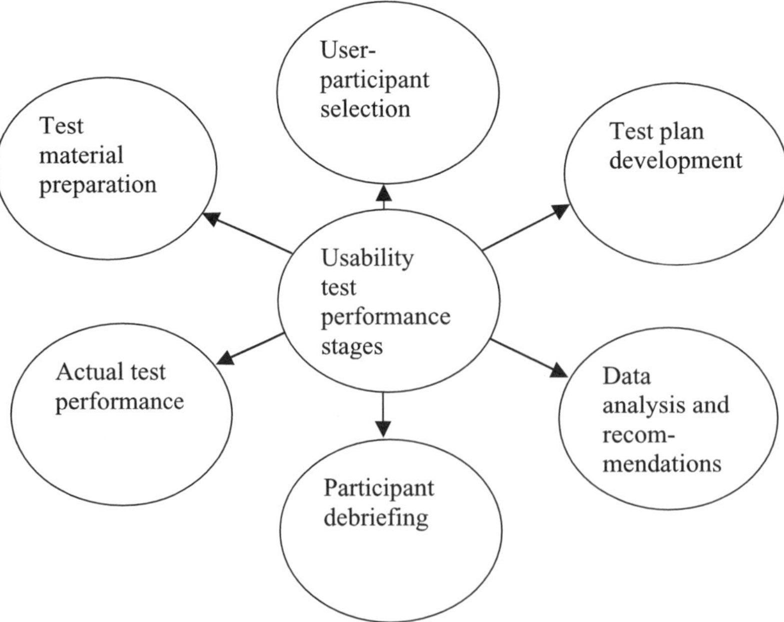


**Fig. 6.3** Stages of performing a usability test

### 5.2. User-Participant Selection

The selection and acquisition of representative user-participants for the product under consideration is a critical component of the usability testing process. The wrong selection of these individuals will lead to end usability-related results of very little value. The starting point in the process of user-participant selection is to establish the user profile of the targeted population by considering factors such as those listed below:

- Skill and experience (general).
- Educational level.
- Skill and experience with respect to the product under consideration.
- The degree of variations in skill, experience, and education levels.
- Demographic information.
- Built-in bias toward the product under consideration (positive or negative).

From time to time, difficulty is experienced during the attempt to acquire information about the end product users for direct or indirect application in user-participant selection. Over the years, professionals working in the field have identified various sources for obtaining such information. Some of the typical ones are presented in Fig. 6.4.

The product functional specification is the blueprint for the item under consideration. More specifically, this document describes all of the intended functions of the product/item, in addition to the tasks the end users will carry out. Usually, a functional specification document contains a comprehensive description of the targeted product users or market for the product.

The structured analysis/marketing studies are usually performed by people such as developers, technical writers, and human factors specialists prior to the start of any design. These studies normally contain information on the skill and knowledge set required to use the product in an effective manner.

The product manager is also a reasonably good source for obtaining the end user-related information, because this individual usually has access to surveys and reports describing the user profile in depth.

The marketing manager is another high-profile individual that often has some in-depth analyses with respect to end users that have not yet been released to the development team members.

Finally, the competitive product analysis group in some organizations performs extensive benchmarking of their own and competitors' products. Consequently, it generates good usability-related data not only on company products but on competitors' as well.

The user-participant selection process requires careful consideration in deciding the number of user-participants to be used in testing a product. Past experiences indicate that this decision depends on many

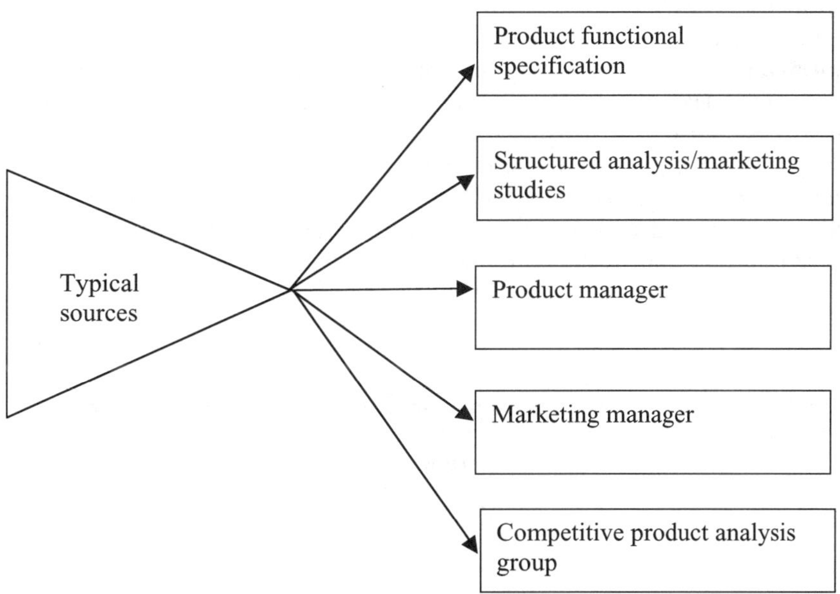


**Fig. 6.4** Typical sources for obtaining information about end product users

factors, including the required degree of confidence in the results, availability of the representative user-participants, the amount of resources available to set up and perform the test, the time required to prepare for the test, and the duration of the testing session.

There are a number of sources for acquiring user-participants for a test. Some of these are as follows [1, 9]:

- In-house lists of existing customers.
- The personnel department of the company.
- Market research firms.
- Employment agencies.
- College campuses.
- Product user groups.

### 5.3. Test Material Preparation

Developing the test materials for use in communicating with the user-participants, collecting data, and satisfying legal requirements is one of the more labour-intensive activities. The specific contents of the required material may vary from one usability test to another. The commonly required material to develop a test may be grouped under two distinct categories: questionnaire-based and non-questionnaire-based. Table 6.1 and 6.2 present the list of items belonging to questionnaire-based and non-questionnaire-based categories, respectively. All 10 items listed in both of the tables are discussed below, separately.

- **Screening questionnaire.** This is the vehicle for qualifying and choosing user-participants to take part in the test. The questionnaire can be simple or complex, depending on the variability and background of the potential user-participants. Usually, the screening questionnaire is presented over the phone to potential user-participants. Some of the useful guidelines for developing a screening questionnaire are as follows:
  - Review the end user/market profile with care.
  - Focus on characteristics unique to the product under consideration.
  - Organize questions in a specific order.
  - Test the questionnaire on colleagues or others, and make revisions as appropriate.
- **Background questionnaire.** This is used to collect historical data about user-participants for understanding their behaviour and performance during the test. More specifically, the questionnaire is composed of a set of questions for obtaining information on user-participants' experience, attitude, and preferences. Some of the guidelines for developing a background questionnaire are as follows:
  - Aim to ascertain all essential information that may affect user-participant performance.
  - Avoid open-ended questions.
  - Test the questionnaire on an individual fitting the end user profile.
- **Pre-test questionnaire.** This questionnaire addresses specific test objectives, including: to ascertain user-participant attitudes and initial impressions about an item's ease of use just before using it; to determine the user-participant's pre-requisite knowledge prior to using the product under consideration; and, to qualify user-participants for inclusion into a specific test group.
- **Post-test questionnaire.** This is used to collect preference-related data from user-participants to have a better understanding of the product's strengths and limitations. Some of the useful guidelines pertaining to the post-test questionnaire are as follows:
  - Ask questions on issues that cannot be observed directly, e.g., feelings, opinions, and suggestions for improvement.

**Table 6.1** Questionnaire-based required material for conducting a usability test

| No. | Required material item |
|---|---|
| 1 | Screening questionnaire |
| 2 | Background questionnaire |
| 3 | Pre-test questionnaire |
| 4 | Post-test questionnaire |

**Table 6.2** Non-questionnaire-based required material for conducting a usability test

| No. | Required material item |
|---|---|
| 1 | Orientation script |
| 2 | Task scenarios |
| 3 | Pre-requisite training materials |
| 4 | Non-disclosure agreement |
| 5 | Data collection instruments |
| 6 | Debriefing topics guide |

- Make use of problem statements from the test plan as the basis for questionnaire contents.
- Emphasize simplicity in designing questions and responses.
- Conduct appropriate pilot tests and refine the questionnaire as necessary.

- **Orientation script.** This is a communication tool for user-participants. Its objective is to put user-participants at ease by describing what will happen during the test session, as well as to set the tone for the session in user-participants' minds. Three important guidelines for developing an orientation script are as follows:
  - Keep the script tone professional and friendly.
  - Aim to read the script to each and every user-participant verbatim.
  - Aim to limit the script length to one or two pages, if possible.
- **Task scenarios.** These are representations of actual work that the user-participants would carry out using the product under consideration. In particular, task scenarios describe items such as work motives, actual data and names rather than generalities, results to be achieved by user-participants, readouts of displays and printouts that will be seen by the user-participants during task performance, and the state of the system at task initiation. Some of the key guidelines for developing task scenarios are as follows:
  - Make every effort to match the task scenarios to user-participants' experiences.
  - Arrange task scenarios in the order they are supposed to be executed.
  - Make every effort to avoid using jargon and cues.
  - Provide scenarios that represent real-life situations as much as possible.
- **Pre-requisite training materials.** These are used to provide training to user-participants prior to the actual usability test for the purpose of raising their skill to some predetermined level. Some of the advantages of the pre-requisite training are: provides a more challenging usability test; permits to test the more obscure functionality of the product; and, forces to understand how someone learns to use the product.
- **Non-disclosure agreement.** This is used to prevent the unauthorized disclosures of proprietary information about product in question that user-participants may encounter during the usability test.
- **Data collection instruments.** These are used to expedite the collection of necessary data to meet the test objective effectively. Here, the main intent is to collect these data simply, concisely, and reliably. The data collected during a usability test falls under two broad categories: preference data and performance data. Prior to collecting data, careful attention should be placed on factors such as data collection mechanism, recording of the data, the test plan problems the data will address, plans to perform data analysis, authority to which data are to be reported, and available resources.
- **Debriefing topics guide.** This provides structure for conducting the debriefing session. Broadly speaking, the debriefing topics guide lists the general topics for discussion. In particular, it may be added that, although the guide suggests a line of questioning, the exact nature of these questions depends on the circumstances surrounding the test session.

### 5.4. Actual Test Performance

This stage is concerned with actually performing the usability test. In order to perform the test effectively, its careful monitoring is essential. Some of the useful guidelines for monitoring the test are as follows:

- Monitor the test session as impartially as possible.
- Consider each new user-participant as an individual.

- Use humor as much as possible to keep the test session participants relaxed.
- Be alert of the effects of one's voice and body language.
- Ensure that all user-participants complete their assigned tasks before proceeding to the next one.
- Avoid rescuing user-participants whenever they struggle.
- Consider using the "thinking aloud" method. This is a simple and useful approach for capturing the thinking of user-participants while working.
- Probe and interact with user-participants as the need arises.
- Some of the useful guidelines in regard to probing are as follows:
  - Avoid asking direct questions as much as possible.
  - Avoid showing surprise.
  - Limit interruptions to short discussions during an ongoing session.
  - Aim to handle one specific issue at a time.
  - Explore opportunities that are useful to understand the rationale for a specific behavior or preference.
  - Focus on the condition the user-participants expected to occur.
- Assist user-participants only in exceptional situations. Here are some conditions under which assistance is justifiable:
  - The user-participant is quite confused or lost.
  - The user-participant is exceptionally frustrated and there is strong likelihood that he/she may give up.
  - The user-participant is uncomfortable in performing the required task.
  - The user-participant's actions cause a failure that requires repair actions.

In the actual process of usability testing, past experiences indicate that it is very advantageous to develop a series of checklists to guide the process. These lists are described in detail in Ref. [14].

### 5.5. Participant Debriefing

This is an important usability test performance stage and is basically concerned with interrogating and reviewing the actions taken by user-participants during the actual performance of the usability test. The basic idea behind debriefing is to clearly understand why each and every error, difficulty, and omission occurred during the test. Thus, the debriefing session with user-participants is the last opportunity to fulfill goals such as these effectively.

The steps of a methodology for conducting a debriefing session after the main performance test are as follows:

- Explore ideas with respect to debriefing while the user-participants complete any post-test questionnaire document.
- Carefully review the completed post-test questionnaire.
- Invite free-wheeling of thoughts from user-participants.
- Start questions from general high-level issues.
- Consider specific issues.
- Review the areas of the post-test questionnaire document completed by the user-participants as those considered to require further exploration.
- Focus on the understanding of problems rather than on problem solving.
- Complete the entire line of questioning prior to inviting input from test observers.
- Leave the door open for further contact with user-participants if the need arises.

### 5.6. Data Analysis and Recommendations

This is the last stage of the usability test performance process and is concerned with analyzing the collected data and making appropriate recommendations. The four major steps involved in analyzing usability test data are shown in Fig. 6.5.

Each of these steps is described in detail below, separately.

- **Compile and Summarize Collected Data.** Compiling is concerned with placing the collected data into a form that permits the understanding of patterns to a certain degree. Usually, the collected data are

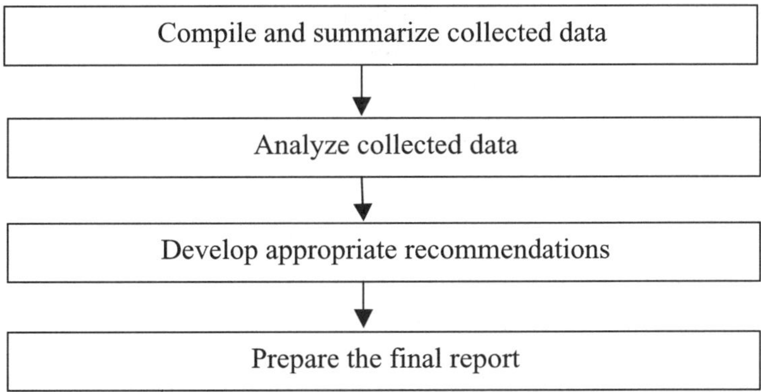


**Fig. 6.5** Major steps involved in analyzing test data

summarized using various mathematical concepts. For example, the performance data in terms of task timings and task accuracy can be summarized as follows:

a) **Task Timings.** These relate to the time user-participants required to accomplish each and every task. Usually, the following statistical concepts are used to summarize task timing data:
   - **Mean time to complete each task (MTTCET).** This is defined by

$$MTTCET = \frac{TT_a}{n} \tag{6.1}$$

   where
   $TT_a$ is the total of all user-participants' completion times for a given task.
   $n$ is the total number of user-participants.

   - **Median time to complete each task.** This is the middle time when all the user-participants' completion times are listed in ascending order for a task.
   - **Range of task completion times.** This is the highest and lowest user-participants' task completion times for each task.
   - **Standard deviation of task completion times.** For each task, the standard deviation ($\sigma$) of user-participants' task completion times is given by

$$\sigma = \frac{\left[\sum t^2 - \frac{\left(\sum t\right)^2}{m}\right]^{1/2}}{m-1} \tag{6.2}$$

   where
   $m$ is the total number of user-participants' completion times for each task.
   $t$ is the time taken by a user-participant to complete a task.

b) **Task accuracy.** There are various ways to determine task accuracy. Three useful indexes that relate to task accuracy are as follows:
   - Percentage of user-participants performing their assigned tasks successfully, irrespective of the time benchmark.
   - Percentage of user-participants performing their assigned tasks successfully within the time benchmark.
   - Percentage of user-participants performing their assigned tasks successfully, irrespective of the time benchmark, including those who required assistance.

- **Analyze collected data.** Once the data are compiled and summarized, the next step is to make sense of the whole thing and draw appropriate conclusions. This requires steps such as those listed below.

  - Identification of tasks that failed to satisfy the set criterion.
  - Identification of user errors and associated difficulties.
  - Identification of the source of each and every error.
  - Prioritization of problems by criticality.
  - Analysis of differences between groups/product versions.
- **Develop appropriate recommendations.** These recommendations are useful to improve the product with respect to usability. In developing such recommendations, careful attention must be paid to factors such as:
  - Emphasis on solutions expected to have the greatest impact on the product.
  - Inclusion of both long-term and short-term recommendations.
  - Identification of areas for further research.
- **Prepare the final report.** The preparation of the final report is equal in importance to any other crucial usability-related tasks. Thus, its effective documentation is absolutely essential. This report usually includes major sections such as executive summary, method, results, findings and recommendations, and appendices.

## 6. TYPICAL USABILITY TEST MONITOR ROLES, CHARACTERISTICS AND PROBLEMS

The role played by the test monitor during usability testing is very critical. More specifically, the effectiveness of the testing is very much subjected to this role. This individual is specifically responsible for all preparations, including test materials, subject arrangements, and coordination of the efforts of other test term members (if any). In particular, during the actual usability test, the monitor's responsibility involves all aspects of administration, including greeting user-participants, collecting all appropriate data, assisting and probing as needs arise, and debriefing all user-participants. In order to perform tasks such as these effectively, a good test monitor should possess characteristics such as those listed in Table 6.3 [1, 15].

Over the years, professionals working in the usability area have identified many problems that are specifically confined to test monitors. Some of the typical ones are as follows:

- Too rigid with the usability test plan.
- Jumping too quickly to conclusions.
- Overly involved with data collection.
- Leading rather than enabling.
- Poor relationships with user-participants.
- Acting too knowledgeable.

All in all, the usability test monitors must avoid problems such as these in order to enhance their performance.

**Table 6.3** Typical characteristics of a good usability test monitor

| No. | Characteristic |
|---|---|
| 1 | Effective knowledge of usability concepts |
| 2 | Quick learning ability |
| 3 | Good listening ability |
| 4 | Flexibility |
| 5 | Empathy |
| 6 | Instant rapport with user-participants |
| 7 | Good humor and memory |
| 8 | Comfortable with ambiguity |
| 9 | Good communication skills |
| 10 | Long attention span |
| 11 | Good emotional control |
| 12 | Tact |

## 7. PROBLEMS

1. What are the principal goals of usability testing?
2. What are the benefits and limitations of usability testing?
3. List at least five important elements of usability testing.
4. Discuss a usability test plan.
5. What are the major cost components of a usability test budget?
6. Discuss the following two types of usability tests:
   - Validation test
   - Assessment test
7. List six important stages of performing a usability test.
8. List at least five important sources for acquiring user-participants for a test.
9. What are the typical characteristics of a good usability test monitor?
10. What are the typical problems that are specifically confined to usability test monitors?

## 8. REFERENCES


1. Rubin, J., Handbook of Usability Testing, John Wiley and Sons, New York, 1994.
2. Gould, J. D., Lewis, C, Designing for Usability: Key Principles and What Designers Think, Communications of the ACM, Vol. 2, No. 3, 1985, pp. 300–311.
3. Dumas, J. S., Stimulating Change Through Usability Testing, SIGCHI Bulletin, 1989, pp. 37–42.
4. Dumas, J. S., Redish, J. C., A Practical Guide to Usability Testing, Ablex Publishing Corporation, Norwood, New Jersey, 1993.
5. Nielsen, J., Usability Engineering, Academic Press, Boston, 1994.
6. Nielsen, J., Usability Testing, in Handbook of Human Factors and Ergonomics, edited by G. Salvendy, John Wiley and Sons, New York, 1997, 1543–1568.
7. Anderson, M., Mahn, G., Usability Testing: What It Is and What It Can Do for Your Documentation, STC Proceedings, 1991, pp. WE48.
8. Asahi, T., Miyai, H., A Usability Testing Method Employing the "Trouble Model", Proceedings of the Human Factors Society Annual Conference, 1990, pp. 1233–1237.
9. Dumas, J. S., Conducting Usability Tests, in Designing User Interfaces for Software, edited by J. S. Dumas, Prentice-Hall, Englewood Cliffs, New Jersey, 1988, pp. 25–30.
10. Mills, C. B., Usability Testing in the Real World, SIGCHI, Bulleting, Vol. 19, No. 1, 1987, pp. 43–46.
11. Mills, C. B., Dye, K. L., Usability Testing: User Reviews, Technical Communication, Vol. 32, No. 4, 1985, pp. 40–44.
12. Winbush, B., McDowell, G., Testing: How to Increase the Usability of Computer Manuals, Technical Communication, Vol. 27, 1980, pp. 20–22.
13. Zirinsky, M., Usability Testing of Documentation, IEEE Transactions on Professional Communication, 1986, pp. 121–125.
14. Brown, D. C., How to Get Started in Usability Testing, STC Proceedings, 1989, pp. 165–168.
15. Dhillon, B. S., Engineering and Technology Management, Artech House, Inc., Boston, 2002.

# Chapter 7

## USER ERRORS

### CONTENTS



## 1. INTRODUCTION

Each year, billions of dollars are spent to produce various types of engineering products, ranging from a simple medical device to a sophisticated computer system. Although most of these products are easy to use, some may require a high level of skills. User expertise in these skills may vary quite considerably.

Today, generally during the design phase of complex engineering products, human factors are given considerable attention. More specifically, the user/operator factor is studied and the interface design is made user-friendly.

Although the history of human factors may be traced back thousands of years, it was not until 1945 when human factors engineering was recognized as a specialized discipline. In 1958, H.L. Williams was probably the first person to recognize that human-element reliability must be included in the overall product-reliability prediction; otherwise, the predicted product reliability would not represent the real picture. This obviously included the user-interface reliability or the occurrence of user/operator errors. Over the years, a large number of publications have appeared on human errors [1]. Many of them were specifically concerned with user/operator error.

This chapter presents various important aspects of user/operator error.

## 2. USER/OPERATOR ERROR FACTS, FIGURES, AND EXAMPLES

Some of the facts, figures, and examples directly or indirectly associated with user/operator error are as follows:

- Human error is considered to contribute or cause up to 90% of accidents generally and in medical devices [2–4].
- A study of electronic equipment reported that 30% of failures were due to operation and maintenance error [5].

ISBN: 1-58883-084-5 / $35.00

- A study of air traffic control system errors revealed that over 90% of the errors were due to operators [6].
- A study reported that operator error accounts for over 50% of all technical medical equipment problems [7].
- In 2001, a total of 1,091 accidents involving steam-heated boilers occurred in the United States [8]. Operator error or poor maintenance was responsible for 37% of these accidents.
- A study revealed that 41% of the 631 accidents involving hot-water boilers in the United States in 2001 were due to either operator error or poor maintenance [8].
- The Center for Devices and Radiological Health (CDRH) of the Food and Drug Administration (FDA) reported that approximately 60% of the deaths or serious injuries associated with medical devices were due to user error [9].
- A patient at a health care facility was seriously injured because an attending nurse incorrectly read the number 7 as 1 [10].
- A fatal radiation-overdose accident involving Therac radiation therapy device was caused by an operator error [11].
- An infant patient became hypoxic during treatment with oxygen because the attending physician set the flow control knob between 1 and 2 litres per minute without realizing the fact that the scale numbers represented discrete, instead of continuous settings [10].
- A patient died because an infusion pump was incorrectly set to deliver 210cc of heparin per hour instead of the ordered 21cc [12].

## 3. OPERATOR ERROR CAUSES AND CLASSIFICATION OF ERRORS MADE IN OPERATING EQUIPMENT

There are many causes that lead to operator/user errors. Some of those are shown in Fig. 7.1 [13].

Human errors made in operating equipment/product can be grouped under two major classifications as follows [14]:

- **Classification I:** Omission errors
- **Classification II:** Commission errors

Omission errors include attention and memory errors. Attention errors are associated to situations requiring operator/user attention. For example, an operator expected to report changes in values displayed on a group of meters. A failure to report a change would be considered an attention error. Memory errors are associated with operator memory. For example, if an operator fails to perform a specified task, it would be a memory error.

Commission errors include identification, interpretation, and operation errors. Identification errors occur when the operator misidentifies objects and treats them as the correct objects. Usually, the frequency of

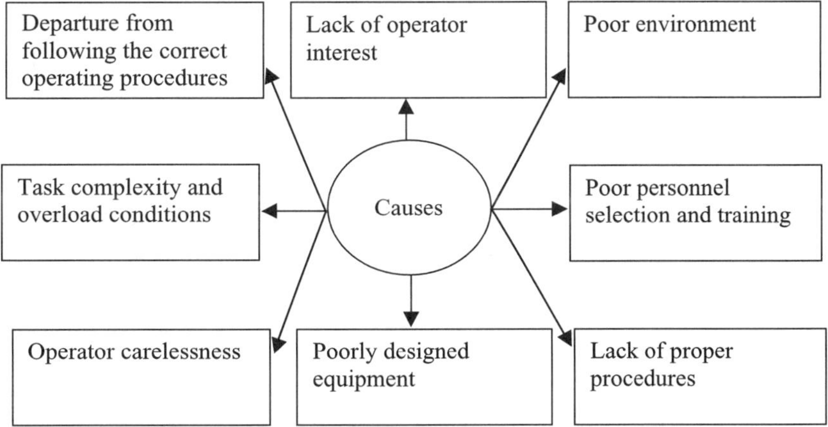


**Fig. 7.1** Causes of operator errors

occurrence of this type of errors is greater than that of any other types of errors [14]. Interpretation errors are associated with the misunderstanding of information and result in the performance of the wrong tasks. Operation errors may be described as those where the control movements are unnecessary to achieve the desired effect.

## 4. CLASSIFICATION* OF USER ERRORS IN HUMAN-COMPUTER INTERACTIVE TASKS AND STRUCTURED QUERY LANGUAGE (SQL) USER ERRORS

User errors occurring in human-computer interactive tasks may be classified under two broad categories: deviations from the correct path in performing the task and failure to complete the task [15]. However, in terms of omission and commission error classifications presented in the previous section, the user errors in computer interactive tasks are discussed below.

Two typical examples of omission errors with respect to computer interactive tasks are as follows [15]:

- Clicking on the paste text icon without copying the text.
- Quitting the application without saving the work.

There are four types of commission errors as shown in Fig. 7.2 [15]. Two of these types with respect to computer interactive tasks are described below and the other two types are considered self-explanatory.

Selection errors occur when users choose a wrong option to carry out a task. Three examples of selection errors are as follows:

- Choosing the incorrect menu option.
- Clicking the incorrect icon.
- Choosing the incorrect option in a dialog box.

Sequence errors occur when users carry out tasks in the wrong order. For example, "merge" feature users had difficulties understanding steps in the procedure, and thus performed them in an incorrect order.

Structured Query Language (SQL) is the standard language used in industry for querying databases [16]. More specifically, database query languages are useful to provide access to essential company/corporate information. In using SQL, users commit many errors [17]. A good understanding of these errors, directly or indirectly, enables designers to create improved interfaces and better documentation.

User errors made in composing SQL queries can be grouped under the following two categories [18].

- **Category I:** Syntactic errors.
- **Category II:** Semantic errors.

Syntactic errors occur when the rules of SQL grammar are violated. They are detected by the database management system's query language interpreter and cause only minor irritants to users. In contrast, seman-

*This classification is basically the same as in the previous section, but here it is specifically concerned with human-computer interactive tasks.

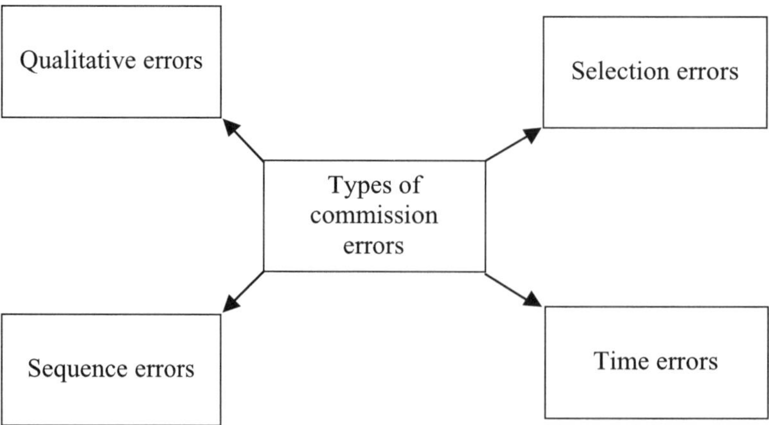


**Fig. 7.2** Types of commission errors

**Table 7.1** User errors in common SQL queries

| No. | Error |
|---|---|
| 1 | Omitting qualifications |
| 2 | Misspellings |
| 3 | Omitting the join clause |
| 4 | Synonyms |
| 5 | Omitting quotes around character data |
| 6 | Omitting the FROM clause |
| 7 | AND/OR difficulties<br>• Use of AND for OR<br>• Improper nesting<br>• Failure to specify column name twice<br>• Using WHERE twice |

tic errors are quite serious and are syntactically correct. More specifically, they are correctly interpreted by the database management system but return information is not intended by users. Semantic errors present problems in the following ways:

- Inability of users committing semantic errors to detect that the retrieved information is incorrect, resulting in wrong and costly business decisions.
- In the event of detecting an error, the user must rectify the query error and resubmit the query. This exercise is a nuisance and a waste of computing and human resources. These errors may affect query users, application programmers and decision makers differently.

Past experiences indicate that users use only about 20% of a language or system to accomplish around 80% of their work. More specifically, as per Ref. [18], only five of the 16 SQL query types account for about 85% of queries written. Furthermore, when users compose these five common SQL queries, they commit SQL errors presented in Table 7.1. These errors, along with their cognitive explanations in parentheses, are omitting qualifications (absence of retrieval cue), misspellings (misperception, imprecise retrieval cues), omitting the join clause (inaccurate procedural knowledge, exceeding working memory's capacity, absence of retrieval cue, and procedural fixedness), synonyms (imprecise retrieval cues), omitting quotes around character data (procedural fixedness, inaccurate procedural knowledge, and absence of retrieval cue), omitting the FROM clause (inaccurate procedural knowledge, absence of retrieval cue, and procedural fixedness), and AND/OR difficulties, e.g., the use of AND for OR (imprecise retrieval cues), improper nesting (inaccurate procedural knowledge), failure to specify column name twice (inaccurate procedural knowledge), and using WHERE twice (procedural fixedness, imprecise retrieval cues, and inaccurate procedural knowledge) [18].

## 5. COMMON MEDICAL DEVICE/EQUIPMENT OPERATOR/USER ERRORS

Each year, billions of dollars are spent to produce various types of medical devices/equipment throughout the world [19]. These devices/equipment range from the relatively simple, such as syringes and catheters, to the most technologically complex and sophisticated, such as computer-controlled diagnostic equipment, and all are subject to user errors. Users of these devices include physicians, nurses, support personnel, and patients. Over the years, various types of studies have been conducted to investigate the occurrence of errors in the use of medical equipment/devices [20]. Some of the common operator/user errors are as follows [21]:

- Inadvertent/untimely/inappropriate activation of controls.
- Incorrect decision-making and actions to critical situations.
- Failure to recognize critical device/equipment output.
- Misinterpretation of vital device/equipment output.
- Errors in properly setting device/equipment parameters.
- Misassembly and inappropriate improvisation.
- Departure from following prescribed procedures and instructions.
- Over-reliance on an equipment's/device's automatic features, capabilities, or alarms.
- Wrong selection of device/equipment with respect to the clinical requirements and objectives.

## 6. METHODS FOR PERFORMING USER ERROR ANALYSIS

Over the years, many methods and techniques have been developed in human factors and human reliability areas to perform various types of analysis [22]. Some of these methods can also be used to conduct user error analysis. This chapter presents four of these methods: failure modes and effect analysis (FMEA), probability tree analysis, fault tree analysis (FTA), and the Markov method.

### 6.1. Failure Modes and Effect Analysis (FMEA)

This is a widely used method in industry to perform reliability and safety studies of engineering systems. The method was developed in the early 1950s to conduct safety analysis of flight control systems [23]. When this method is extended to classify each failure effect according its severity, it is called failure mode effects and criticality analysis (FMECA) [24].

FMEA can also be used to perform user/operator error analysis and is described in more detail in Chapter 5.

### 6.2. Probability Tree Analysis

This method was developed to perform human reliability analysis in the nuclear power industry [25]. The method is quite useful to perform task analysis by diagrammatically representing critical human actions and other events associated with the system under consideration. Diagrammatic representations result in a probability tree structure. The method is described in detail in Chapter 5.

### 6.3. Fault Tree Analysis (FTA)

This is one of the most widely used methods in industry to perform reliability and safety analyses of engineering systems. The method was developed in the early 1960s to perform safety analysis of the Minuteman Launch Control System [26]. FTA is described in detail in Chapter 5. It can be used to perform user/operator error analysis. The following example demonstrates FTA application to perform user error analysis:

---

**Example 7.1**

Assume that user error X occurs if error A, B, C, or D occurs. In turn, error A occurs if errors $a_1$ and $a_2$ occur, error B occurs if error $b_1$ or $b_2$ occurs, and error C occurs if errors $c_1$ and $c_2$ occur.

Develop a fault tree for the top event "user error X occurrence" by using Fig. 5.4 symbols.

---

A fault tree for the example using Fig. 5.4 symbols is shown in Fig. 7.3.

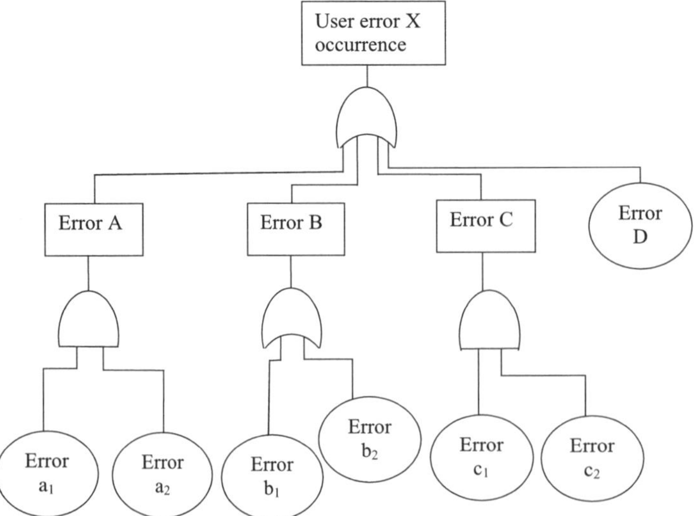


**Fig. 7.3** A fault tree for the top event: user error X occurrence

**Example 7.2**

Assume that in Example 7.1, the probabilities of occurrence of errors D, $a_1$, $a_2$, $b_1$, $b_2$, $c_1$, and $c_2$ are 0.02, 0.03, 0.04, 0.05, 0.06, 0.07, and 0.08, respectively. Calculate the probability of occurrence of user error X by using the Fig. 7.3 fault tree.

By substituting the given data into Equation (5.6), error A occurrence probability is

$$P(A) = (0.03)(0.04) = 0.0012$$

where

$P(A)$ is the probability of occurrence of error $A$.

Using the given data values in Equation (5.7) yields

$$P(B) = 1 - (1 - 0.05)(1 - 0.06) \\ = 0.107$$

where

$P(B)$ is the probability of occurrence of error $B$.

By inserting the specified data values into Equation (5.6), we get

$$P(C) = (0.07)(0.08) \\ = 0.0056$$

where

$P(C)$ is the probability of occurrence of error $C$.

Using all of the above calculated values and given data in Equation (5.7), we obtain the following value for the probability of occurrence of user error X:

$$P(X) = 1 - (1 - 0.0012)(1 - 0.107)(1 - 0.0056)(1 - 0.02) \\ = 0.1308$$

Thus, the probability of occurrence of user error X is 0.1308. Fig. 7.4 shows the Fig. 7.3 fault tree with given and calculated error occurrence probability values.

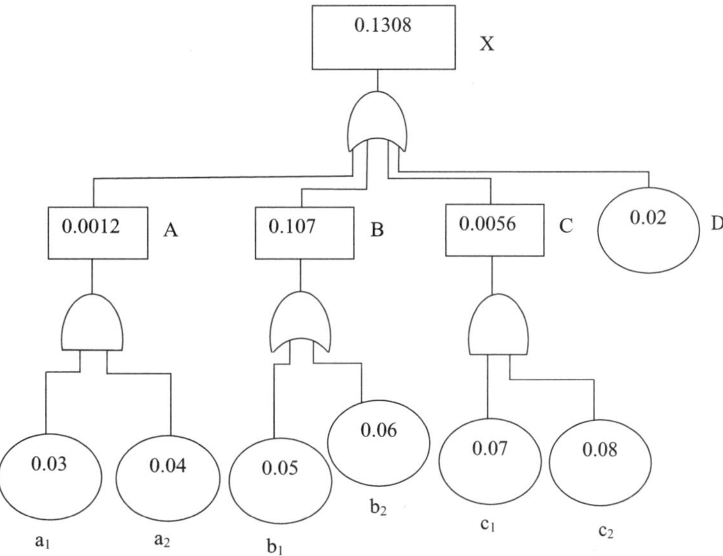


**Fig. 7.4** A fault tree with error occurrence probability values (plain and subscripted letters in the diagram denote errors)

### 6.4. Markov Method

This is a powerful tool widely used in industry to perform various types of reliability and availability studies, particularly when the repairability of items is considered. The method is named after Russian mathematician, Andrei Andreyevich Markov (1856–1922), and its applicability to human reliability problems is demonstrated in Ref. [22]. This method is described in Chapter 5. Its applicability to user error-related problems is demonstrated through the following mathematical models:

#### Model I

This model represents an operator/user performing a time-continuous task. The operator/user can commit an error in performing the task. The model state space diagram is shown in Fig. 7.5. The numerals in boxes denote corresponding state. For a given user error rate, the Markov method can be used to predict operator/user reliability at time *t*, operator/user probability of committing an error at time *t*, and operator/user mean time to error.

The following symbols are associated with the model:

- $i$ is the operator/user state; for $i = 1$ means operator/user is performing the time-continuous task normally, $i = 2$ means operator/user committed an error.
- $P_i(t)$ is the probability that the operator/user is in state $i$ at time $t$; for $i = 1, 2$.
- $\lambda_0$ is the operator/user constant error rate.
- $s$ is the Laplace transform variable.

With the aid of the Markov method, we write down the following differential equations for Fig. 7.5 diagram:

$$\frac{dP_0(t)}{dt} + \lambda_0 P_0(t) = 0 \quad \textbf{(7.1)}$$

$$\frac{dP_1(t)}{dt} - \lambda_0 P_1(t) = 0 \quad \textbf{(7.2)}$$

At time $t = 0$, $P_0(0) = 1$, and $P_1(0) = 0$.

Solving Equations (1) and (2) with the aid of the Laplace transforms, we get:

$$P_0(s) = \frac{1}{s + \lambda_0} \quad \textbf{(7.3)}$$

$$P_1(s) = \frac{\lambda_0}{s(s + \lambda_0)} \quad \textbf{(7.4)}$$

where

- $P_0(s)$ is the Laplace transform of the probability that the operator/user is in state 0.
- $P_1(s)$ is the Laplace transform of the probability that the operator/user is in state 1.

After taking the inverse Laplace transform of Equations (7.3) and (7.4), we obtain:

$$P_1(t) = 1 - e^{-\lambda_0 t} \quad \textbf{(7.5)}$$

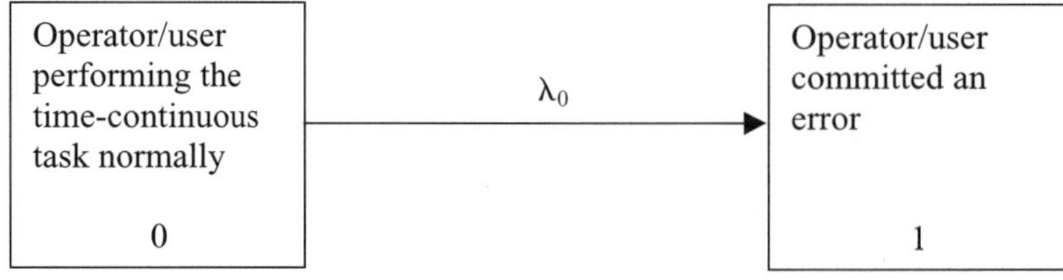


**Fig. 7.5** State space diagram for an operator/user performing a task

$$P_1(t) = 1 - e^{-\lambda_0 t} \tag{7.6}$$

Equation (7.5) is operator/user reliability at time. By integrating Equation (7.5) over the time interval [0, ∞], we get:

$$\begin{aligned} MTTOE &= \int_0^\infty e^{-\lambda_0 t}\, dt \\ &= \frac{1}{\lambda_0} \end{aligned} \tag{7.7}$$

where

*MTTOE* is the mean time to operator/user error.

---

**Example 7.3**

An operator/user is performing a certain task and his/her error rate is 0.0005 errors/hour. Calculate the following:

- Probability of the operator/user committing an error during a 500-hour mission.
- Mean time to operator/user error.

By substituting the specified data values into Equations (7.6) and (7.7), we get

$$\begin{aligned} P_1(500) &= 1 - e^{-(0.0005)(500)} \\ &= 0.2211 \end{aligned}$$

and

$$\begin{aligned} MTTOE &= \frac{1}{(0.0005)} \\ &= 2{,}000 \text{ hours} \end{aligned}$$

Thus, the probability of the operator/user committing an error during a 500-hour mission and mean time to operator/user error are 0.2211 and 2,000 hours, respectively.

---

### Model II

This model is basically same as Model I, but instead of the operator/user performing his/her task in constant environment, the operator/user performs tasks in fluctuating environments (i.e., normal and abnormal or normal and stress) [27]. The operator/user can commit an error in either normal or abnormal environment. Moreover, the operator/user error rate may increase quite significantly under a stressful environment.

The model state space diagram is shown in Fig. 7.6. The numerals in boxes and circles denote corresponding states.

The following assumptions are associated with this model:

- Operator/user error rates are constant.
- Operator/user errors occur independently.
- Operator/user performs a time-continuous task.
- The rate of environment change from normal to stressful (or abnormal) and vice-versa is constant.

The following symbols are associated with the model:

$i$ is the operator/user state; for $i$ = 1 means operator/user is performing the time-continuous task correctly in normal environment, $i$ = 1 means operator/user committed an error in normal environment, $i$ = 2 means operator/user is performing the time-continuous task correctly in abnormal environment, $i$ = 3 means operator/user committed an error in abnormal environment.

$P_i(t)$ is the probability that the operator/user is in state $i$ at time $t$; for $i$ = 0,1, 2, 3.

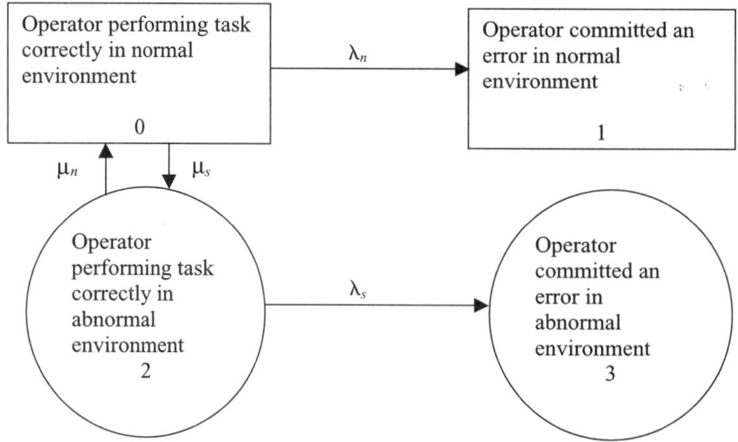


**Fig. 7.6** State space diagram for the operator/user performing under fluctuating environments

$\lambda_n$ is the operator/user error rate when working in normal environment.
$\mu_n$ is the transition rate from abnormal environment to normal environment.
$\mu_s$ is the transition rate from normal environment to abnormal or stressful environment.
$\lambda_s$ is the operator/user error rate when working in abnormal environment.

With the aid of the Markov method, we write down the following system of differential equations for Fig. 7.6:

$$\frac{dP_0(t)}{dt} + (\lambda_n + \mu_s)P_0(t) = \mu_n P_2(t) \tag{7.8}$$

$$\frac{dP_2(t)}{dt} + (\lambda_s + \mu_n)P_2(t) = \mu_s P_0(t) \tag{7.9}$$

$$\frac{dP_1(t)}{dt} = \lambda_n P_0(t) \tag{7.10}$$

$$\frac{dP_3(t)}{dt} = \lambda_s P_2(t) \tag{7.11}$$

At time $t = 0$, $P_0(0) = 1$, and $P_1(0) = P_2(0) = P_3(0) = 0$.
By solving Equations (7.8)–(7.11), we get:

$$P_0(t) = (w_1 - w_2)^{-1}\left[(w_2 + \lambda_s + \mu_n)e^{w_2 t} - (w_1 + \lambda_s + \mu_n)e^{w_1 t}\right] \tag{7.12}$$

where

$$w_1 = \frac{-c_1 + \sqrt{c_1^2 - 4c_2}}{2}$$

$$w_2 = \frac{-c_1 - \sqrt{c_1^2 - 4c_2}}{2}$$

$$c_1 = \lambda_n + \lambda_s + \mu_n + \mu_s$$

$$c_2 = \lambda_n(\lambda_s + \mu_n) + \mu_s \lambda_s$$

$$P_1(t) = c_4 + c_5\, e^{w_2 t} - c_6\, e^{w_1 t} \tag{7.13}$$

where

$$c_3 = \frac{1}{w_1 - w_2}$$

$$c_4 = \frac{\lambda_n(\lambda_s + \mu_n)}{w_1 + w_2}$$

$$c_5 = c_3(\lambda_n + c_4 w_1)$$

$$c_6 = c_3(\lambda_n + c_4 w_2)$$

$$P_2(t) = \mu_s c_3(e^{w_2 t} - e^{w_1 t}) \quad \textbf{(7.14)}$$

$$P_3(t) = c_7\left[(1 + c_3)(w_1 e^{w_2 t} - w_2 e^{w_1 t})\right] \quad \textbf{(7.15)}$$

where

$c_7 = (\lambda_s \mu_s)/w_1 w_2$

The operator/user reliability in alternating environment is:

$$R_0(t) = P_0(t) + P_2(t) \quad \textbf{(7.16)}$$

where

$R_0(t)$ is the operator/user reliability at time t in alternating environment.

The mean time to operator/user error is given by:

$$\begin{aligned} MTTOE_f &= \int_0^\infty R_0(t)dt \\ &= (\lambda_s + \mu_s + \mu_n)/c_2 \end{aligned} \quad \textbf{(7.17)}$$

where

$MTTOE_f$ is the mean time to operator/user error in fluctuating environment.

---

**Example 7.4**

An operator/user performs his/her time-continuous tasks in fluctuating (i.e., normal and abnormal) environments with constant error rates 0.0004 errors/hour and 0.0008 errors/hour, respectively. The constant transition rates from normal to abnormal or stressful environment and vice-versa are 0.0009/hour and 0.0007/hour, respectively. Calculate the mean time to operator/user error.

Substituting the given data values into Equation (7.17) yields:

$$\begin{aligned} MTTOE_f &= \frac{(0.0008 + 0.0009 + 0.0007)}{(0.0004)(0.0008 + 0.0007) + (0.0009)(0.0008)} \\ &= 1818.18 \text{ hours} \end{aligned}$$

This means the operator/user will commit an error after every 1,818.18 hours.

---

## 7. PROBLEMS

1. Write an essay on user errors.
2. Discuss at least five important facts, figures, and examples concerned with user/operator errors.
3. What are the important causes for the occurrence of operator errors?

4. Discuss the following:
   - Omission errors
   - Commission errors
5. Discuss user errors in human-computer interactive tasks.
6. List at least seven user errors in common structured query language (SQL) queries.
7. Discuss common medical device/equipment user errors.
8. Discuss at least four methods that can be used to perform user error analysis.
9. An operator is performing a certain time-continuous task and his/her constant error rate is 0.0002 errors per hour. Calculate the probability of the operator committing an error during a 400-hour mission.
10. Using Equations (7.12) and (7.14), prove that the mean time to operator/user error in fluctuating environment is given by Equation (7.17).

## 8. REFERENCES


1. Dhillon, B. S., Human Errors: A Review, Microelectronics and Reliability, Vol. 29, 1989, pp. 299–304.
2. Maddox, M. E., Designing Medical Devices to Minimize Human Error, Med. Dev. Diag. Ind. Mag., Vol. 19, No.5, 1997, pp. 166–180.
3. Nobel, J. L., Medical Device Failures and Adverse Effects, Pediat. Emerg. Care, Vol. 7, 1991, pp. 120–123.
4. Bogner, M. S., Medical Devices and Human Error, Human Performance in Automated Systems: Current Research and Trends, edited by M. Moulona and R. Parasuraman, Lawrence Erlbaum Associates Publishers, Hillsdale, New Jersey, 1994, pp. 64–67.
5. AMCP 706-134, Maintainability Guide for Design, U.S. Army Material Command, Department of the Army, Washington, D.C., 1972.
6. Kenney, G. C., Spahn, M. J., Amato, R. A., The Human Element in Air Traffic Control: Observations and Analysis of Performance of Controllers and Supervisors in Providing Air Traffic Control Separation Services, METREK Div., MITRE Corp., Report No. MTR-7655, December 1977.
7. Dhillon, B. S., Reliability Technology in Health Care Systems, Proceedings of the IASTED International Symposium on Computers and Advanced Technology in Medicine, Health Care, and Bio-Engineering, 1990, pp. 84–87.
8. The National Board of Boiler and Pressure Vessel Inspectors, 1055 Crupper Avenue, Columbus, Ohio, USA.
9. Bogner, M. S., Medical Devices: A New Frontier for Human Factors, CSERIAC Gateway, Vol. IV, No. 1, 1993, pp. 12–14.
10. Sawyer, D., Do It by Design: Introduction to Human Factors in Medical Devices, Center for Devices and Radiological Health (CDRH), Food and Drug Administration (FDA), Washington, D.C., 1996.
11. Casey, S., Set Phasers on Stun: and other True Tales of Design Technology and Human Error, Aegean Inc., Santa Barbara, California, 1993.
12. Brueley, M. E., Ergonomics and Error: Who is Responsible?, Proceedings of the First Symposium on Human Factors in Medical Devices, 1989, pp. 6–10.
13. Meister, D., Human Factors: Theory and Practice, John Wiley and Sons, New York, 1976.
14. Lincoln, Human Factors in the Attainment of Reliability, IRE Transactions on Reliability and Quality Control, Vol. 11, 1962, pp. 97–103.
15. Vora, P., Classifying User Errors in Human-Computer Interactive Tasks, Common Ground (Usability Professionals Association), Vol. 5, No. 2, 1995, pp. 15–16.
16. Date, C. J., A Guide to the SQL Standard, Addison-Wesley, Reading, Massachusetts, 1989.
17. Ogden, W. C., Korenstein, R., Smelcer, J. B., An Intelligent Front-End for SQL, Technical Report, IBM General Products Division, San Jose, California, 1986.
18. Smelcer, J. B., User Errors in Database Query Composition, International Journal of Human-Computer Studies, Vol. 42, 1995, pp. 353–381.
19. Murray, K., Canada's Medical Device Industry Faces Cost Pressures, Regulatory Reform, Med. Dev. Diag. Ind. Mag., Vol. 19, No. 8, 1997, pp. 30–39.
20. Hyman, W. A., Errors in the Use of Medical Equipment, in Human Error in Medicine, edited by M. S. Bogner, Lawrence Erlbaum Associates, Publishers, Hillsdale, New Jersey, 1994, pp. 327–347.
21. Hyman, W. A., Human Factors in Medical Devices, Encyclopaedia of Medical Devices and Instrumentation, edited by J. G. Webster, Vol. 3, John Wiley and Sons, New York, 1988, pp. 1542–1553.
22. Dhillon, B. S., Human Reliability with Human Factors, Pergamon Press, Inc., New York, 1986.
23. Arnzen, H. E., Failure Mode and Effect Analysis: A Powerful Engineering Tool for Component and System Optimization, Report No. 347.40.00.00-K4-05 (C5776), GIDEP Operations Center, Corona, California, 1966.

24. Jordan, W. E., Failure Modes, Effects and Criticality Analyses, Proceedings of the Annual Reliability and Maintainability Symposium, 1972, pp. 30–37.
25. Swain, A. D., A Method for Performing a Human-Factors Reliability Analysis, Report No. SCR-685, Sandia Corporation, Albuquerque, New Mexico, August 1963.
26. Haasl, D. F., Advanced Concepts in Fault Tree Analysis, Proceedings of the System Safety Symposium, 1965. Available from the University of Washington Library, Seattle.
27. Dhillon, B. S., Stochastic Models for Predicting Human Reliability, Microelectronics and Reliability, Vol. 25, 1985, pp. 729–752.

# Chapter 8

## USABILITY COSTING

### CONTENTS



## 1. INTRODUCTION

Cost is an important factor in product development and other issues and it generally varies with time. In fact, the concept of time as a value of money is not that new, it began over 2,500 years ago in Babylon [1]. In those days, interest was paid on borrowed commodities in form of grain or other means. In modern times, the development cost of an item not only varies with time, but also due to various other factors, including usability. More specifically, effectively usable products are not produced by accident, but through a careful consideration given to human factors or usability during their design and development. This costs money.

The cost of each usability engineering activity depends on factors such as [2]:

- Scope of the product under consideration.
- Functionality range.
- Number of users to be studied.
- Number of scenarios to be studied.
- Skill and experience of the usability specialists.

Although the cost of usability engineering activities may become a significant component of product development cost, past experiences indicate that each dollar spent by a manufacturer in developing the usability of a product returns \$10–\$100 in benefits, in addition to winning customer satisfaction and continued business. This chapter presents various important aspects of usability costing.

## 2. USABILITY COSTING-RELATED FACTS AND FIGURES

Some of the direct or indirect usability costing-related facts and figures are as follows:

- It costs approximately \$100 billion annually in lost productivity to American businesses because office workers "futz" with their machines an average of 5.1 hours per week [3].

ISBN: 1-58883-084-5 / \$35.00

- A study reported that the training time for new users of a standard personal computer was around 21 hours as opposed to only 11 hours for users of a more usable computer machine [4].
- A study revealed that the average cost of end-user computing across 18 major Australian companies for supporting a single workstation was approximately $10,000 (Australian) [5]. At least 50% of this cost accounted for "hidden" support (i.e., productivity lost because users stopped their ongoing tasks to help each other with computer-related problems).
- A study conducted by American Airlines revealed that catching a usability-related problem early in the design could decrease the cost of rectifying it by 60% to 90% [6].
- A study reported that human factors improvements resulted in $2.5 million savings in training cost [7].
- A study reported that in ergonomically designed office environments, absenteeism dropped from 4% to slightly greater than 1% [8].
- An Australian insurance company spent around $100,000 (Australian) on a usability project concerned with redesigning its application forms to make customer errors less likely and saved $536,023 (Australian) annually [9].
- A study revealed that an $800,000 investment in reducing the human factors errors for the Line of Sight Forward Heavy resulted in an $80 million cost savings [10, 11].
- A study reported that a $300,000 investment in reducing human factors-related errors for the Pedestal Mounted Stinger resulted in a $61 million cost savings [10, 11].
- It is estimated that approximately 63% of all software projects exceed their cost estimates due to factors such as frequent requests for changes from users, users' lack of understanding of their own requirements, overlooked tasks, and inadequate user-analyst communication and understanding [12].
- It is estimated that roughly 80% of software maintenance cost is due to unforeseen/unmet user needs [13].
- A study reported that in 1991, design changes to usability work at IBM resulted in an average decrease of 9.6 minutes per task, with projected internal savings of $6.8 million [14].

## 3. USABILITY ENGINEERING ACTIVITIES AND COSTS

There is a wide range of activities that may be employed in developing a usability engineering product: end user requirements definition, benchmark studies, user profile definition, surveys and questionnaires, usability objectives specification, style guide development, focus groups, heuristic evaluations, task analysis, prototype redesign, design walkthroughs, usability tests (laboratory or field), studies of end user work context, paper-and-pencil simulation testing, thinking-aloud studies, prototype development (high or low fidelity), and initial design development [15, 16].

The cost of usability engineering for a given product includes costs for one or more of the above activities. Usually, no more than six of these activities are completed for any one project/product [15]. Moreover, usability engineering-related work on a project is tailored to factors such as the project's requirements, time frame, and resources. Nonetheless, two important guidelines for calculating the cost of usability engineering activities are as follows [15]:

- Ensure that in personnel cost, costs for all development team support, other support, or contract services, and all costs associated with participants, are included.
- Ensure that cost is prorated on the basis of the number of usability tests to be performed for a specified period, when a permanent usability laboratory is built.

## 4. COST-BENEFIT ANALYSIS OF A USABILITY STUDY

The performance of a usability study costs money, but in turn, it generates various benefits. More specifically, the main objective of a usability investment is to accrue maximum returns. Many times a quick estimate of cost and benefit is made before making a decision for performing a usability study. Nonetheless, the following example demonstrates the estimation of the cost of a usability study along with the estimation of savings after doing the usability study.

**Example 8.1**

Estimate the cost and savings of a usability study concerned with reducing the number of support center calls and improving productivity of employees by producing a better engineered, usability tested product. Assume fictitious values as applicable.

### Usability Study Cost Estimation

In the calculation of this cost, consideration is given to in-house usability staff, the product of time spent by staff (i.e., usability individuals and developers) and wage rate (same units), and additional variable costs associated with subcontractors, usability laboratory rentals, travels, etc.

For fictitious values of amount and time, the cost of the in-house usability staff is estimated as follows:

- Annual loaded headcount amount (this includes salary, benefits, and cost with respect to vacation leave, office space, telephones, and equipment) = $ 151,200
- Annual hours per work year (i.e., 35 hours per week times 48 weeks) = 1,680
- Hourly wage rate = $151,200/1,680 = $90 per hour
- Amount of time spent on usability test by usability specialist for activities such as planning, analysis, implementation, and recommendations = 140 hours
- Amount of time spent by interface designer on redesign activity = 50 hours
- Total amount of time spent by development engineers on usability-related activities = 20 hours
- Total cost of the in-house usability staff = (140 + 50 + 20) (90) = $18,900

Similarly, for fictitious values of amount, the cost of a fully equipped usability laboratory is estimated as follows:

- Total cost of usability participant recruiting at $200 per participant for 8 participants = $1,600
- Total amount of participant compensation at $70 per participant for 8 participants = $560
- Total cost of videotapes at $10 per tape for 8 tapes = $80
- Total amount or percentage of laboratory and equipment cost (more specifically, this is amortized cost of laboratory per hour (i.e., $250/hour) for 25 hours) = $6,250
- Total cost of the fully equipped usability laboratory = (1,600 + 560 + 80 + 6,250) = $8,490

Thus, the total usability study cost is:

$$\begin{aligned} TUSC &= \$18{,}900 + \$8{,}490 \\ &= \$27{,}390 \end{aligned}$$

where

TUSC is the total usability study cost.

### Usability Study Savings Estimation

These savings are the result of performing the usability study. Two major components of these savings are savings in support call cost and in-house increase in productivity.

For fictitious values, total savings in support call cost is estimated as follows:

- Average cost of a support call = $100
- Total number of product version A units sold = 150,000
- Total number of support calls = 300,000
- Total cost of support calls due to usability-related problems = (300,000 × $100) = $30 million
- Number of support calls per product version A unit sold = (300,000/150,000) = 2
- After the application of usability engineering, the total number of product version B units sold = 400,000
- Total number of support calls with version B units = 100,000
- Total cost of support calls due to usability-related problems with version B units (100,000×$100) = $10 million

- Number of support calls per version B unit sold = 100,000/400,000 = 0.25
- Reduction on support calls (i.e., from version A to version B) = 2–0.25 = 1.75
- Total savings in support call cost due to improved usability = (1.75) (400,000) ($100) = $70 million

Similarly, for fictitious values, the annual improvement (in dollars) in in-house productivity is estimated as follows:

- Improvement in time, because of the usability action, in performing Task X = 2 minutes
- Number of times task X is performed per day = 4
- Total number of users performing task X = 300
- Hourly wage of users, including overheads = $70
- Total user time saved per day because of the usability action = (2) (4) (300) = 2,400 minutes or 40 hours
- Total amount of money saved per day because of the usability action = (70) (40) = $2,800
- Total annual amount of money saved through increased productivity = ($2,800) (240 work days/year) = $672,000 or $0.672 million
- Total savings after performing the usability study = 70 + 0.672 = $70.672 million

Thus, the total usability study estimated cost and the total estimated savings after performing the usability study are $27,390 and $70.672 million, respectively.

## 5. COST OF IGNORING USABILITY AND COST EFFECTIVENESS OF USABILITY EVALUATION METHODS

Past experiences indicate that over the years, many products from usability or human factors standpoints were poorly designed, but put on the market and have subsequently requited in substantial costs to manufacturers in term of reduced sale, increased customer dissatisfaction, tarnished corporate image, and so on. Two examples of scenarios such as these are discussed in Ref. [7]. Nonetheless, some of the possible principal costs of ignoring product usability are shown in Fig. 8.1 [17]. Poor sale cost is associated with dissatisfied users not buying the product in future, even if they are made aware of improvements in the usability of a next version of the product. Moreover, it is estimated that a dissatisfied user roughly influences ten others to avoid purchasing the brand [17].

Customer support cost pertains to a customer hotline telephone service provided by product manufacturers for individuals having problems using the product. Products that are difficult to use generate greater user/customer requests for help, and thus require more people to handle the customers/users. In turn, the manufacturer incurs greater customer support costs. The cost of a tarnished corporate image is rather difficult to estimate. It occurs when customers/users avoid purchasing not only the current or improved usability version of the product, but also other products manufactured by the same company.

User error cost is associated with the users of professional products making errors. These errors lead to a decrease in their productivity. More specifically, the probability of user error increases significantly if products are difficult to use; thus higher user error cost. Poor productivity cost is associated with the additional time spent by users of professional products that are difficult to use. Cost due to poor productivity increases significantly if these products are used by a large number of users on a daily basis.

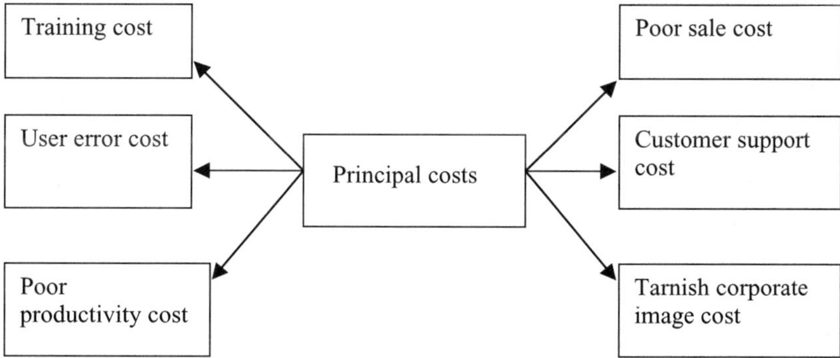


**Fig. 8.1** Principal costs of ignoring product usability

**Table 8.1** Classifications of usability evaluation methods with respect to cost

| No. | Low cost methods | Medium cost methods | High cost methods |
|---|---|---|---|
| 1 | Heuristic evaluation | Remote testing | Focus group |
| 2 | — | Interviews | Coaching method |
| 3 | — | Cognitive walkthroughs | Thinking-aloud protocol |
| 4 | — | Questionnaires | Shadowing method |
| 5 | — | Logging actual use | Question-asking protocol |
| 6 | — | Field observation | Teaching method |
| 7 | — | Scenario-based checklists | Co-discovery learning |
| 8 | — | Proactive field study | Performance measurement |
| 9 | — | Feature inspection | Pluralistic walkthroughs |
| 10 | — | — | Retrospective testing |

Training cost is associated with the training of users when the product or system is first introduced. It increases quite significantly for products that are difficult to use. In the case of user-friendly products, this cost could be very little or negligible.

In usability studies, a variety of usability evaluation methods are used. The cost of employing a usability evaluation method is subject to factors such as:

- Number of involved individuals (i.e., users, developers, usability experts, etc.).
- Amount of time required for collecting and analyzing data.
- Degree of need for coordination (i.e., whether a given approach needs all the participants to be present together).

Usability evaluation methods may be classified under three distinct cost categories (i.e., low cost, medium cost, and high cost) as presented in Table 8.1.

## 6. MODELS FOR ESTIMATING VARIOUS TYPES OF USABILITY ENGINEERING COSTS

This section presents four mathematical models that can be used to estimate various types of usability engineering-related costs.

### 6.1. Model I

This model can be used to approximate the usability engineering cost of a product when usability engineering cost data are available for similar products of different capacities. The usability engineering cost of the desired product can be estimated by using the following relationship [18]:

$$UC_d = UC_0\left[\frac{K_d}{K_0}\right]^{\alpha} \tag{8.1}$$

where

$UC_d$ is the usability engineering cost of the desired product.

$K_d$ is the capacity of the desired product.

$UC_0$ is the known usability engineering cost of a similar product, of capacity $K_d$.

$\alpha$ is known as the cost-capacity factor and its value varies for different products. In circumstances where no data are available, it is generally reasonable to assume the value of $\alpha$ to be 0.6.

---

**Example 8.2**

Assume that the usability engineering cost of a 20 GB hard drive personal computer is \$200. Estimate the usability engineering cost of a similar 80 GB hard drive personal computer, if the value of the cost-capacity factor is 0.7.

Substituting the given data into Equation (1) yields:

$$UC_d = (200)\left[\frac{80}{20}\right]^{0.7} = \$527.8$$

It means the usability engineering cost of the 80 GB hard drive personal computer is $527.80.

### 6.2. Model II

This model can be used to estimate the labor cost of correcting usability-related problems associated with a product for a given period. Thus, the labor cost of correcting product usability-related problems is expressed by

$$C_u = (SOH)(LC)\left[\frac{MTTCPUP}{MTTPUP}\right] \quad \textbf{(8.2)}$$

where

- $C_u$ is the labor cost of correcting product usability-related problems for a given period.
- $LC$ is the hourly labor cost of correcting usability problems.
- $SOH$ is the total product scheduled operating hours for a specified period.
- $MTTPUP$ is the mean time to product usability problems expressed in hours.
- $MTTCPUP$ is the mean time to correct product usability problems expressed in hours.

**Example 8.3**

A product is scheduled to be used for 3,000 hours annually. Its mean time to usability problems and mean time to correct usability problems are 100 hours and 5 hours, respectively. The labor cost of correcting usability problems is $50 per hour. Calculate the annual labor cost of correcting usability problems.

Substituting the given data values into Equation (2) yields:

$$C_u = (3{,}000)(50)\left[\frac{5}{100}\right]$$
$$= \$7{,}500$$

Thus, the annual labor cost of correcting usability problems is $7,500.

### 6.3. Model III

This model can used to estimate user error cost over a product's life span. This cost is expressed by:

$$UELCC = UEOC + UECC \quad \textbf{(8.3)}$$

where

- $UELCC$ is the user error life cycle cost of a product.
- $UEOC$ is the user error occurrence cost associated with the product. This is the cost of productivity loss due to user errors over the product life span.
- $UECC$ is the user error correction cost. This is associated with time spent correcting user errors over the product life span.

### 6.4. Model IV

This model can be used to estimate the usability engineering life cycle cost of a product. The usability engineering life cycle cost is expressed by:

$$LCC_{ue} = AC_{ue} + OC_{ue} \quad \textbf{(8.4)}$$

where

$LCC_{ue}$ is the usability engineering life cycle cost of a product.

$AC_{ue}$ is the usability engineering acquisition cost of the product. This cost is an element of a product's procurement cost and is concerned with usability engineering or human factors activities during the product design and development phase.

$OC_{ue}$ is the usability engineering ownership cost of the product. This cost is a component of a product's ownership (i.e., operational phase) cost and is associated with usability-related problems during the product operational phase.

## 7. PROBLEMS

1. Write an essay on usability costing.
2. Discuss direct or indirect usability costing-related facts and figures.
3. List at least ten major usability engineering activities with respect to cost.
4. Discuss the cost-benefit analysis of a usability study.
5. List and discuss principal costs of ignoring product usability.
6. List at least six medium cost usability evaluation methods.
7. Discuss user error cost.
8. What is usability engineering life cycle cost?
9. Assume that the usability engineering cost of a 10 GB hard drive personal computer is $150. Estimate the usability engineering cost of a similar 100 GB hard drive computer, if the value of the cost-capacity factor is 0.6.
10. A product is scheduled to be used for 4,000 hours annually. Its mean time to usability problems and mean time to correct usability problems are 200 hours and 4 hours, respectively. The labor cost of correcting usability problems is $40 per hour. Calculate the annual labor cost of correcting usability problems.

## 8. REFERENCES


1. Paul-DeGarmo, E., Canada, J. R., Sullivan, W. G., Engineering Economy, Macmillan Publishing Co., Inc., New York, 1979.
2. Rosson, M. B., Carroll, J. M., Usability Engineering: Scenario-Based Development of Human-Computer Interaction, Morgan Kaufmann Publishers, San Francisco, 2002.
3. SBT Accounting Systems, 1997, Westlake Consulting Company, Inc., 5444 Westheimer, Unit 1510, Houston, Texas.
4. Nielson, J., Usability Engineering, Academic Press, Inc., Boston, 1993.
5. Ko, C., Hurley, M., Managing End-User Computing, Information Management and Computer Security, Vol. 3, No. 3, 1995, pp. 3–6.
6. Laplante, A., Put to the Test, Computerworld, Vol. 27, July 27, 1992, pp. 75–77.
7. Chapanis, A., The Business Case for Human Factors in informatics, in Human Factors for Informatics Usability, edited by B. Shackel and S. J. Richardson, Cambridge University Press, London, 1991, pp. 39–71.
8. Schneider, M. F., Why Ergonomics Can No Longer be Ignored, Office Administration and Automation, Vol. 46, No. 7, 1985, pp. 26–29.
9. Fisher, Pl, Sless, D., Information Design Methods and Productivity in the Insurance Industry, Information Design Journal, Vol. 6, No. 2, 1990, pp. 103–129.
10. Booher, H. R., Rouse, W. B., MANPRINT as the Competitive Edge, in MANPRINT: An Approach to Systems Integration, edited by H. R. Booher, Van Nostrand Reinhold Company, New York, 1990, pp. 230–245.
11. Rouse, W. B., Boff, K. R., Assessing Cost/Benefits of Human Factors, in Handbook of Human Factors and Ergonomics, edited by G. Salvendy, John Wiley and Sons, New York, 1997, pp. 1617–1633.
12. Lederer, A. L., Prasad, J., Nine Management Guidelines For Better Cost Estimating, Communications of the ACM, Vol. 35, No. 2, 1992, pp. 51–59.
13. Pressman, R. S., Software Engineering: A Practioner's Approach, McGraw- Hill Book Company, New York, 1992.

14. Karat, C. M., Cost-Benefit Analysis of Usability Engineering Techniques, Proceedings of the Human Factor Society 34th annual Meeting, 1990, pp. 839–843.
15. Karat, C. M., A Business Case Approach to Usability Cost Justification, in Cost-Justifying Usability, edited by R. G. Bias and D. J. Mayhew, Morgan Kaufmann Publishers, San Francisco, 1994.
16. Mantei, M. M., Teorey, T. J., Cost-Benefit Analysis for Incorporating Human Factors in the Software Lifecycle, Communications of the ACM, Vol. 31, No. 4, 1988, pp. 428–439.
17. Keinonen, T., Mattelmaki, T., Soosalu, M., Sade, S., Usability Design Methods, Technical Report, Department of Product and Strategic Design, University of Art and Design, Helsinki, Finland, 1997.
18. Dieter, G. E., Engineering Design, McGraw Hill Book Company, New York, 1983, pp. 324–366.

# Chapter 9

# MATHEMATICAL MODELS FOR USABILITY ASSURANCE

## CONTENTS



## 1. INTRODUCTION

Mathematical models are widely used in engineering to study various types of phenomena. In these models, the parts of an item are denoted by idealized elements containing all the representative characteristics of the real-life parts and whose behaviour may be described by equations. The realism of a given mathematical model depends very much on assumptions imposed upon it.

Ever since the inception of the human factors field, many useful mathematical models and formulas have been developed for various purposes, including determining the length of rest period for a given task, the visual angle for the legibility of a written message, the character height for effective legibility at a given distance, the channel capacity required for transmitting perfectly intelligible speech, and the effectiveness of written text [1–3]. Although, the effectiveness of these models may vary from one application to another, they can be used to improve usability of engineering products.

This chapter presents a number of mathematical models or formulas considered useful to enhance usability of engineering products directly or indirectly.

ISBN: 1-58883-084-5 / $35.00

## 2. MOVEMENT TIME ESTIMATION MODEL

This is also known as Fitts' Law model and is used to estimate movement distance. With respect to usability, the model allows the designer to estimate the amount of time needed to position, say, a computer cursor on a displayed object/item as a function of the distance to be traveled and the size of the item/object to be chosen [4]. The movement time is expressed by [5–6]:

$$T_m = \alpha + \beta\theta \tag{9.1}$$

where
- $T_m$ is the movement time.
- $\theta$ is the index of difficulty.
- $\alpha$ is the intercept constant.
- $\beta$ is the slope constant.

In turn, $\theta$, is defined by:

$$\theta = \log_2 [2d/w_t] \tag{9.2}$$

where
- $d$ is the distance from the starting point to the center of the target.
- $w_t$ is the width of the target.

To correct for the tendency of Equation (9.1) to predict lower than observed movement times at rather low values of $\theta$, Ref. [7] proposed the following equation for estimating $\theta$:

$$\theta = \log_2 [d/(w_t + 0.5)] \tag{9.3}$$

## 3. VISUAL ANGLE ESTIMATION MODEL

This model is useful to estimate the value of the visual angle. Information about the visual angle is important because message legibility depends on the visual angle subtended at the individual's eye by the viewed object. The visual angle can be estimated by using the following equation [8]:

$$VA = [(60)(57.3)(H_C)/D_t] \tag{9.4}$$

where
- $VA$ is the visual angle expressed in minutes.
- $H_C$ is the height of the character or the target diameter.
- $D_t$ is the total distance from the target.

The constant values given in Equation (9.4) can only be used for angles less than ten degrees (i.e., 10°).

## 4. CHARACTER HEIGHT ESTIMATION MODEL I

This model is useful to estimate optimum character height. An equation for the model is expressed in terms of the following three factors:

- The viewing distance under consideration.
- The viewing conditions under consideration.
- The importance of numbers.

Thus, the optimum height of a letter is expressed by [9–10]:

$$H_o = K(VD) + CF_{vC} + CF_{Cn} \tag{9.5}$$

**Table 9.1** $CF_{vC}$ values under various conditions

| No. | Condition | $CF_{vC}$ values |
|---|---|---|
| 1 | Favourable reading conditions under high ambient illumination. | 0.06 |
| 2 | Unfavourable reading conditions under high ambient illumination. | 0.16 |
| 3 | Favourable reading conditions under low ambient illumination. | 0.16 |
| 4 | Unfavourable reading conditions under low ambient illumination. | 0.26 |

where

- $H_o$ is the optimum height of a letter expressed in inches.
- $K$ is the constant with specified value of 0.0022.
- $VD$ is the viewing distance expressed in inches.
- $CF_{Cn}$ is the correction factor for the criticality of the number. Its two values are 0.075 (when the number is very critical) and 0.0 (when the number is other than very critical).
- $CF_{vC}$ is the correction factor for viewing condition. Its values under various conditions are given in Table 9.1.

## 5. READABILITY INDEX MODELS

Today, the term "readability" is used as a measurable aspect of writing and, over the years, many readability formulas, indexes, or models have been developed [11]. The purpose of these models is to provide feedback to writers whether their written material is easy or difficult to understand. The models are based on the relationships between the average word length (i.e., number of syllables per word) and average length of the sentence (i.e., number of words per sentence). Two main drawbacks of the readability index models are as follows:

- Difficult to compare the results of difference readability index models because of inconsistencies in the model equations.
- The factors of content difficulty are ignored.

Two widely used readability index models are presented below.

### 5.1. Fog Index Model

This model was developed by Robert Gunning and it estimates a figure, X, representing the number of years of education needed to understand the material under consideration [11]. The Fog index, X, is defined by:

$$X = \lambda(W_m + W_s) \tag{9.6}$$

where

- $\lambda$ is the constant with the specified value of 0.4.
- $W_m$ is the mean number of words per sentence. Past experiences indicate that for accuracy, one should aim to take a sample of at least 100 words in length.
- $W_s$ is the total number of words having more than two syllables per 100 words. In estimating the value of $W_s$, avoid counting items such as combinations of short and easy words, capitalized words, and three-syllable verbs with added "es" or "ed".

### 5.2. Flesch-Kincaïd Index Model

This model is used to estimate the education grade level required to comprehend the written material under consideration. The following equation is used to estimate the education grade level required [12–13]:

$$EGL = \theta_1 MW + \theta_2 MS - \theta_3 \tag{9.7}$$

where

$EGL$ is the education grade level.
$\theta_i$ is the ith constant; for $i = 1$ ($\theta_1 = 0.39$), $i = 2$ ($\theta_2 = 11.8$), $i = 3$ ($\theta_3 = 15.59$).
$MS$ is the mean number of syllables per word.
$MW$ is the mean number of words per sentence.

---

**Example 9.1**

A piece of written material has an average of 10 words per sentence. In turn, there are an average of two syllables per word. Calculate the readability grade level of the materials by using Equation (9.7).

By substituting the specified data values into Equation (9.7) we get:

$$\begin{aligned} EGL &= (0.39)(10) + (11.8)(2) - 15.59 \\ &\approx 12 \end{aligned}$$

It means approximately 12 years of education are required to understand the written material.

---

## 6. CHANNEL CAPACITY ESTIMATION MODEL

This model is also known as the Shannon formula and is used to estimate the channel capacity required to transmit perfectly intelligible speech. The model uses the following equation to estimate channel capacity [14–15]:

$$CC = BW \log_2 \left[ 1 + \frac{SP}{NP} \right] \tag{9.8}$$

where

$CC$ is the channel capacity expressed in bits per second.
$SP$ is the signal power.
$NP$ is the noise power.
$BW$ is the channel bandwidth.

## 7. ILLUMINANCE ESTIMATION MODEL

This model is used to estimate the value of illuminance. The model defines illuminance as follows [2]:

$$I = \frac{LI}{D} \tag{9.9}$$

where

$I$ is the illuminance or illumination expressed in lux. A lux is one lumen per square meter.
$D$ is the distance from the light source expressed in meters.
$LI$ is the luminous intensity expressed in candelas.

## 8. REFLECTANCE ESTIMATION MODEL

This model is used to estimate the value of reflectance. The model defines reflectance as follows:

$$REF = \frac{LUM}{I} \tag{9.10}$$

where

*REF* is the reflectance.
*LUM* is the luminance expressed in candelas per square meter. Luminance is the amount of light per unit area departing from a surface.

For a perfect reflecting surface, the value of REF is equal to unity.

## 9. BRIGHTNESS CONTRAST ESTIMATION MODEL

This model is used to calculate brightness contrast. It expresses the brightness contrast as follows [16]:

$$BC = (LBA - LDA)(100)/LBA \quad \textbf{(9.11)}$$

where

*BC* is the brightness contrast.
*LDA* is the luminance of the darker of two contrasting areas.
*LBA* is the luminance of the brighter of two contrasting areas.

---

**Example 9.2**

Assume that a certain type of paper has a reflectance of 85% and the reflectance of print on the paper is 15%. Calculate the value of the brightness contrast by using Equation (9.11).

In this example, the specified values of LDA and LBA are 15% and 85%, respectively. By substituting these two values into Equation (9.11), we get:

$$\begin{aligned} BC &= (85 - 15)(100)/85 \\ &\approx 82\% \end{aligned}$$

The value of the brightness contrast is approximately 82%.

---

## 10. GLARE CONSTANT ESTIMATION MODEL

This model is concerned with estimating the value of the glare constant. The model defines the glare constant as follows [6]:

$$GC = \theta_s^a \, SL^b/\theta_v^2 \, L_g \quad \textbf{(9.12)}$$

where

*GC* is the glare constant.
a and b are the constants with specified values of 0.8 and 1.6, respectively.
*SL* is the source luminance.
$\theta_s$ is the solid angle subtended by the source at the eye.
$L_g$ is the luminance of the general background.
$\theta_v$ is the angle between the viewing direction and the glare source direction.

GC = 35 is considered as the boundary of "just acceptable" glare and GC = 150 is the boundary of "just uncomfortable" glare.

## 11. CONTROL-DISPLAY RATIO ESTIMATION MODEL

This model can be used to estimate the value of the control-display (C/D) ratio, which is the ratio of the control movement distance to that of the display moving element. Although the ratio is applicable only to continuous controls, it is considered to be a key design factor that affects operator performance. Some studies indicate that, in comparison to a poor control-display ratio, a good control-display ratio can save from 0.5 to

5 seconds in positioning time. A control-display (C/D) ratio for controls involving considerable rotational movement and that affect linear displays is defined as follows [14]:

$$(C/D) = A/DM \quad \textbf{(9.13)}$$

$$A \equiv 2\pi(LAL)(CAM)/360 \quad \textbf{(9.14)}$$

where

*DM* is the display movement.
*CAM* is the control angular movement expressed in degrees.
*LAL* is the lever arm length.

The range of optimum values of the control-display ratio for ball levers and knobs, respectively, are as follows:

- 2.5:1 to 4:1
- 0.2 to 0.8

## 12. CHARACTER HEIGHT ESTIMATION MODEL II

This is another model that can be used to estimate optimum character height. The model uses the following equation to estimate optimum character height [2, 17]:

$$H_{nC} = H_S D_r / B \quad \textbf{(9.15)}$$

where

$H_S$ is the standard or recommended character height at a viewing distance of 28 inches.
$B$ is the constant with a specified value of 28 inches. The specific reason for setting B equal to 28 inches is that, for a comfortable arm reach to conduct control and adjustment-related tasks, normally the instrument panels are located at a viewing distance of 28 inches.
$D_r$ is the required distance expressed in inches.
$H_{nC}$ is the estimated height of a character at $D_r$ expressed in inches.

---

**Example 9.3**

It is estimated that a meter of a newly installed system has to be read from a distance of 36 inches. The recommended numeral height at a viewing distance of 28 inches at low luminance is 0.35 inches. Calculate the numeral height for the 36-inch viewing distance.

Inserting the given data values into Equation (9.15), we get:

$$\begin{aligned} H_{nC} &= (0.35)(36)/28 \\ &= 0.45 \text{ inches} \end{aligned}$$

It means the numeral height for the 36-inch viewing distance is 0.45 inches.

---

## 13. SOUND LOUDNESS ESTIMATION MODEL

This model is internationally agreed to estimate the loudness of a sound. The model uses the following equation to estimate the sound loudness [4]:

$$L_s = 2^N \quad \textbf{(9.16)}$$

$$N \equiv (L_p - 40)/10 \quad \textbf{(9.17)}$$

where

$L_s$ is the loudness of a sound expressed in sones.

$L_p$ is the loudness level of a sound in phons.

## 14. HUMAN PERFORMANCE ESTIMATION MODEL

This model is concerned with measuring inspector performance associated with inspection tasks. The inspector/human performance is expressed by [17]:

$$IP = \frac{TIT}{N - M} \tag{9.18}$$

where

$IP$ is the inspector performance expressed in minutes per correct inspection.

$TIT$ is the total inspection time.

$N$ is the total number of patterns inspected.

$M$ is the total number of inspector errors.

Finally, it may be said that this index (i.e., Equation (9.18)) is an average over a specified trial.

## 15. REST PERIOD ESTIMATION MODEL

Humans perform various types of tasks in their day-to-day work activity. The length of the rest period required may vary from one type of task to another. In the design of engineering systems for use by humans, the factor of user rest period requirement must be taken into consideration carefully for their ultimate effectiveness.

This model uses the following equation to estimate the required duration of a rest period for a given task [18]:

$$RRP = T_t\,(AEE - SEE)/(AEE - \lambda) \tag{9.19}$$

where

$RRP$ is the required rest period expressed in minutes.

$\lambda$ is the resting level. Its approximate value is 1.5 kilocalories per minute.

$T_t$ is the total working time expressed in minutes.

$AEE$ is the average energy expenditure per minute of work expressed in kilocalories.

$SEE$ is the standard energy expenditure expressed in kilocalories per minute. Its value may be taken as 5 kilocalories per minute when no data are available.

---

**Example 9.4**

Assume that a person is performing a computer-interactive task for 60 minutes and his/her average energy expenditure is approximately 4 kilocalories per minute. Calculate the length of the required rest period if the standard energy expenditure is 3 kilocalories per minute.

By substituting the specified data values into Equation (9.19) we get:

$$\begin{aligned} RRP &= (60)(4 - 3)/(4 - 1.5) \\ &= 24 \text{ minutes} \end{aligned}$$

It means the length of the required rest period is 24 minutes.

---

## 16. NOISE REDUCTION ESTIMATION MODEL

Noise could be a pressing problem in industrial and other settings and it must be reduced to an acceptable level. Its reduction is not only dependent on the transmission loss alone, but also on the absorption properties of walls in the receiving room and the area of the wall transmitting sound. Consequently, the following equation can be used to calculate total noise reduction [19]:

$$NR_t = \lambda + \log (AP_w / A_s) \quad \textbf{(9.20)}$$

where

$NR_t$ is the total noise reduction.

$A_s$ is the total area of wall transmitting sound expressed in $ft^2$.

$AP_w$ is the absorption properties of walls in the noise receiving room.

$\lambda$ is the transmission loss of materials of the varying thicknesses expressed in decibels. More specifically, $\lambda$ is defined by the following equation:

$$\lambda = A_S / (\theta_1 A_{S1} + \theta_2 A_{S2} + \ldots + \theta_n A_{Sn}) \quad \textbf{(9.21)}$$

where

$\theta_j$ is the $j$th transmission coefficient of material in question; for $j = 1, 2, 3, n$.

$A_{Sj}$ is the $j$th corresponding areas of the material in question; for $j = 1, 2, 3, n$.

## 17. SOUND-PRESSURE LEVEL ESTIMATION MODEL

This model is concerned with measuring the level of sound/noise intensity in terms of decibels. The bel (B), the basic unit of measurement is named after Alexander Graham Bell, the inventor of the telephone. The sound-pressure level (SPL), in decibels (dB) is defined as follows [2, 3]:

$$SPL(dB) = 10 \log_{10} \left[ \frac{SP^2}{SP_r^2} \right] \quad \textbf{(9.22)}$$

where

$SP^2$ is the sound pressure squared of the sound to be measured.

$SP_r^2$ is the standard reference sound pressure squared denoting zero decibels. More specifically, $SP_r$ is the faintest 1,000-Hertz (Hz) tone that an average person can hear.

## 18. LIFTING LOAD ESTIMATION MODEL

This model is concerned with estimating the maximum lifting load for an individual. The following equation is used to estimate the maximum lifting load [20]:

$$L_m = C(IBMS) \quad \textbf{(9.23)}$$

where

$L_m$ is the maximum lifting load.

$C$ is a constant and its specified values for males and females are 1.1 and 0.95, respectively.

$IBMS$ is the isometric back muscle strength.

## 19. PROBLEMS

1. Write an essay on readability indexes.
2. A technical report has an average of nine words per sentence. In turn, there are an average of three syllables per words. Calculate the readability grade level of the material by using the Flesch-Kincaid index model.
3. What is the Shannon formula?
4. Define the following:
   - Reflectance
   - Illuminance

5. A certain type of paper has a reflectance of 80% and the reflectance of print on the paper is 20%. Calculate the value of the brightness contrast by using Equation (9.11).
6. What is the control-display ratio?
7. Define glare constant.
8. Who developed the Fog readability index?
9. It is estimated that a meter of a newly installed system has to be read from a viewing distance of 30 inches. The recommended numeral height at a viewing distance of 28 inches at low luminance is 0.35 inches. Calculate the numeral height for a 30-inch viewing distance.
10. Assume that a person is performing a computer-interactive task for 50 minutes and his/her average energy expenditure is approximately 3 kilocalories per minute. Calculate the length of the required rest period if the standard energy expenditure is 2 kilocalories per minute.

## 20. REFERENCES


1. Salvendy, G., editor, Handbook of Human Factors and Ergonomics, John Wiley and Sons, New York, 1997.
2. McCormick, E. J., Sanders, M. S., Human Factors in Engineering Design, McGraw-Hill Book Company, New York, 1982.
3. Adams, J. A., Human Factors Engineering, MacMillan Publishing Company, New York, 1989.
4. McMillan, G. R., Eggleston, R. G., Anderson, T. R., Non-conventional Controls, in Handbook Human of Factors and Ergonomics, edited by G. Salvendy, John Wiley and Sons, New York, 1997, pp. 729–827.
5. Fitts, P. M., The Information Capacity of the Human Motor System in Controlling the Amplitude of Movement, Journal of Experimental Psychology, Vol. 47, 1954, pp. 381–391.
6. Fitts, P. M., Peterson, J. R., Information Capacity of Discrete Motor Responses, Journal of Experimental Psychology, Vol. 67, 1964, pp. 103–112.
7. Welford, A. T., The Measurement of Sensory-Motor Performance: Survey and Reappraisal of Twelve years' Progress, Ergonomics, Vol. 3, 1960, pp. 189–230.
8. Grethers, W. F., Baker, C. A., Visual Presentation of Information in Human Engineering, in Guide to Equipment Design, edited by H. P. Van Cott and R. G. Kinkade, U.S. Government Printing Office, Washington, D.C., 1972, Chapter 3.
9. Peters, G. A., Adams, B. B., Three Criteria for Readable Panel Markings, Product Engineering, Vol. 30, 1959, pp. 55–57.
10. Oborne, D. J., Ergonomics at Work, John Wiley and Sons, New York, 1982.
11. Gunning, R., The Technique of Clear Writing, McGraw-Hill Book Company, New York, 1952.
12. Flesch, R. F., How to readability, Harper, Inc., New York, 1951.
13. Burnett, R. E., Technical Communication, Wadsworth Publishing Company, Belmont, California, 1990.
14. Human Engineering Guide to Equipment Design, sponsored by Joint Army-Navy-Air Force Steering Committee, John Wiley and Sons, New York, 1972.
15. Shannon, C. E., A Mathematical Theory of Communication, Bell Syst. Tech. J., Vol. 27, 1948, pp. 379–423 and 623–656.
16. Dhillon, B. S., Human Reliability: with Human Factors, Pergamon Press, Inc., New York, 1986.
17. Drury, C. G., Fox, J. G., Editors, Human Reliability in Quality Control, John Wiley and Sons, New York, 1975.
18. Murrell, K. F. H., Human Performance in Industry, Reinhold Publishing Company, New York, 1965.
19. Dale-Huchingson, R., New Horizons for Human Factors in Design, McGraw-Hill Book Company, New York, 1981.
20. Poulsen, E., Jorgensen, K., Back Muscle Strength, Lifting and Stooped Working Postures, Applied Ergonomics, Vol. 2, 1971, pp. 133–137.

# Chapter 10

## SOFTWARE USABILITY

### CONTENTS



### 1. INTRODUCTION

Today, much more money is spent developing computer software than hardware, in comparison to the first generation computers. For example, in the 1950s, the software element accounted for around 20% of the total computer cost and in the mid-1980s, the percentage increased to approximately 90% [1]. Today, the annual software industry in the United States alone is worth at least $300 billion [2].

In the 1970s, it became clear to computer professionals that user interface design would become an important element of software engineering. With the increasing development of software for interactive use, attention to the preferences and requirements of end users intensified significantly. Today, the user interface is frequently the single most important factor in the success or failure of a software project. Moreover, it is estimated that approximately 50% to 80% of all source code developed accounts for user interface [3].

User-friendly software enables its users to conduct their tasks intuitively and easily, and it significantly supports rapid learning and high skill retention of the involved manpower. It may be said that in today's competitive environment, usability is not a luxury but a basic ingredient in software systems, as productivity and comfort of users relate directly to it.

Software usability may simply be defined as quality in software use [4]. More specifically, it means that how easy the software is to learn and use, how productively its potential users will be able to work, and how much support the users will require [5]. This chapter presents various important aspects of software usability.

### 2. NEED FOR CONSIDERING USABILITY DURING SOFTWARE DEVELOPMENT AND HUMAN-COMPUTER INTERFACE PRINCIPLES

In order to produce user effective software products, careful consideration to usability during their development is absolutely necessary. Nonetheless, some of the specific factors that dictate the need for considering usability during software development are shown in Fig. 10.1. Mixed users mean that the users of the software products could be professionals or non-professionals with limited or no computer skills. Competition means that failure to address usability issues can result in the loss of market share, should competitors

ISBN: 1-58883-084-5 / $35.00

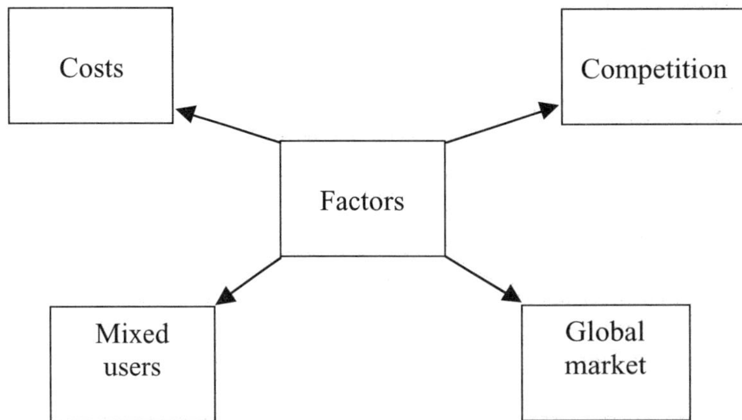


**Fig. 10.1** Important factors for considering usability during software development

release their software products with higher usability. Cost means that poor usability software products reduce user productivity and increase developer cost in terms of hotlines, customer support service, and so forth. Global market means that software products usually cover a global market with varying cultures, language proficiencies, and so on.

Five fundamental principles of the human-computer interface with respect to software are as follows [6]:

- The software must satisfy all task requirements.
- The software ergonomics principles must be applied effectively, in particular, to human data processing.
- The software must be user friendly, easy to use, and adoptable to levels of experience or knowledge of operators (as appropriate).
- The system must be able to display information in a format and at a pace adopted for operators.
- The software system must provide an effective feedback to all users/workers with respect to their performance.

## 3. SOFTWARE USABILITY ENGINEERING PROCESS AND STEPS FOR IMPROVING USABILITY OF A SOFTWARE PRODUCT

The software usability engineering process may be viewed differently from one organization to another. A typical process used in product development is basically composed of the following three principle activities in parallel [7]:

- **Visiting customers to understand their requirements.** This is an important activity for gaining insight into customers' current experience with a system. Data on users' experience are collected primarily through contextual interviews (i.e., all relevant interviews are conducted while users carry out their work activity). During these interviews, users are asked about their work, about their system interfaces, about their perception of the system, and so on. One significant advantage of these contextual interviews is that they rapidly generate large amounts of data.

  Nonetheless, it should be noted that different users in different contexts have different user requirements. The aspects of a user's context that influence the usability of a system for each individual include physical workplace environment, interaction with other software systems, type of work being performed, social situation, and organization culture.
- **Establishing an operational usability specification for the software system under consideration.** Usability specification may simply be described as a measurable definition of usability that is shared by all involved individuals. This is based on the understanding of the user requirements, competitive analysis, and the resources required to produce the software system. Two important points to note in establishing the usability specification are as follows [7]:
  - Failure to clearly understand the user requirements prior to developing a specification can result in a specification document that does not reflect users' requirements.

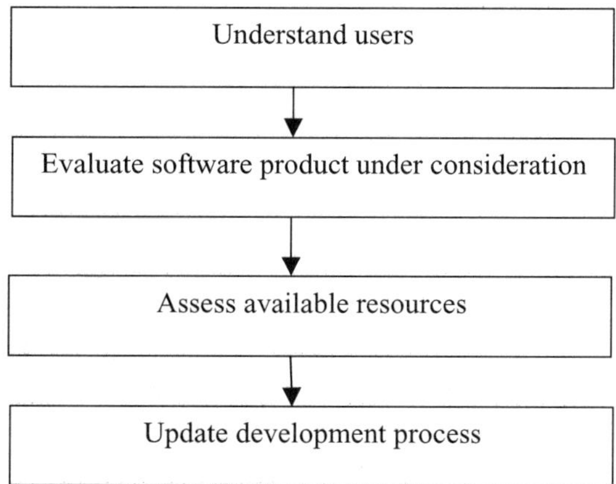


**Fig. 10.2** Steps for improving software usability

- Individuals involved with the development of the usability specification must evaluate it on a continuous basis during the development process and make appropriate changes to reflect up-to-date information on users' requirements.

- **Adopting an evolutionary delivery approach to software system development.** This means first building a small subset of the software system and then "growing" it throughout the development process and continuously studying users as the system evolves. Furthermore, it can be said that evolutionary delivery exploits, rather than overlooks, the dynamic nature of software needs.

  Some of the methods employed to improve software system usability during the stages of evolutionary delivery are building and testing early prototypes, collecting user feedback during early field tests, instrumenting a system to collect usage data, and analyzing the impact of design solutions [8–10].

  An important advantage of the evolutionary delivery approach is that it helps to build the project team members' shared understanding of the software system's user-interface design.

Software usability can be improved by following the four steps* shown in Fig. 10.2 [11]. These are: understand users, evaluate software product under consideration, assess available resources, and update development process. Each of these steps is described in detail below [11].

### Understand Users

Two items critical to understanding users are the user profile and the product usage environment. The user profile helps to focus design efforts on real issues concerning users and avoid wasting time and resources on sideline issues. However, the user profile requires the collection of information on users' needs, demographics, interests, etc. More specifically, on items such as computer expertise, profession and education, age range, content knowledge, percentage of male versus female, and cultural background. Some of the good sources for obtaining this type of information are marketing and sales staff, product registrations, training and customer support staff, and market research data including surveys and studies.

The product usage environment helps to understand the range of factors that impact product utilization. Some of the environmental factors taken into consideration are system configuration, degree of privacy/noise levels, type of network and security, location (i.e., mobile, home, office, etc.), connection speed (if applicable), browser types and settings (if applicable), and other applications likely to be in use. All in all, a clear understanding of factors such as these can be useful to help plan the degree of field research that is to be conducted.

### Evaluate Software Product Under Consideration

Usability is relative. It basically means that a software product may be easy and usable for one type of user, but confusing or intuitive for other users. Thus, an evaluation from the users' perspective means determin-

*These steps may overlap with the software usability engineering process.

ing if there is a good fit between product and all of its potential users. Past experiences indicate that the greater the amount of available data, the more effective the evaluations.

Nonetheless, an effective approach to product (i.e., software) evaluations includes actions such as analyzing the available user data, conducting user field research (i.e., ethnographic research), testing usability, and determining the user-product "fit".

The action "analyzing the available user data" calls for a careful analysis of the data obtained from sources such as technical support, sales support, marketing people, and customer service.

The action "conducting user field research" calls for collecting first-hand information on the functioning of the software with its intended users. The best time to conduct user field research is at the start of the design or redesign project (i.e., to collect information on current usage and on unresolved user requirements).

The action "testing usability" is concerned with determining exactly where users are experiencing difficulties in using the software product. Usability testing can be carried out in many different settings, including a designated area within the company establishment, on customer premises, and an outside market research facility.

The action "determining the user-product 'fit'" is basically concerned with assessing and prioritizing usability issues. This should be performed by keeping in mind that the most critical interface elements that impact the success of a software include graphic quality, screen language clarity, personalization of content, underlying behavioural metaphor, error handling, presentation performance, and intuitive navigation.

### Assess Available Resources

This is concerned with assessing the existing resources and the ability of the involved individuals to execute the software design/redesign project. The diverse talents and capabilities of the design team should also be taken into consideration. For example, it could be extremely useful to produce quality screen writing, graphic (page) design, and interaction design.

Also, it is important to assess the existence of user advocacy among the design team members because it is a key for producing a successful user-centered design. All in all, there should be at least one individual among the team members to advocate user point of views.

### Update Development Process

In order to maintain an effective usability process for future releases or uses, a careful evaluation of the overall development cycle is needed. The critical components of a user-centered design process are keeping satisfactory design-related documentation, establishing methods of tracking results, and establishing cycles of user feedback. Nonetheless, it must be noted that the documents such as those listed below are useful for software design/redesign work [11].

- Functional specification
- User interface specification
- Application specification
- Flowcharts
- Marketing (business) requirements

All in all, the careful evaluation of documents such as these is an important step toward updating the development process.

## 4. SOFTWARE USABILITY INSPECTION METHODS

These methods are the evaluation methods in which usability experts examine the software user interface. There are many usability inspection/evaluation methods [12]. Five of these methods are as follows [11]:

- Pluralistic walkthrough
- Heuristic evaluation
- Guidelines checklists
- Standards inspection
- Cognitive walkthrough

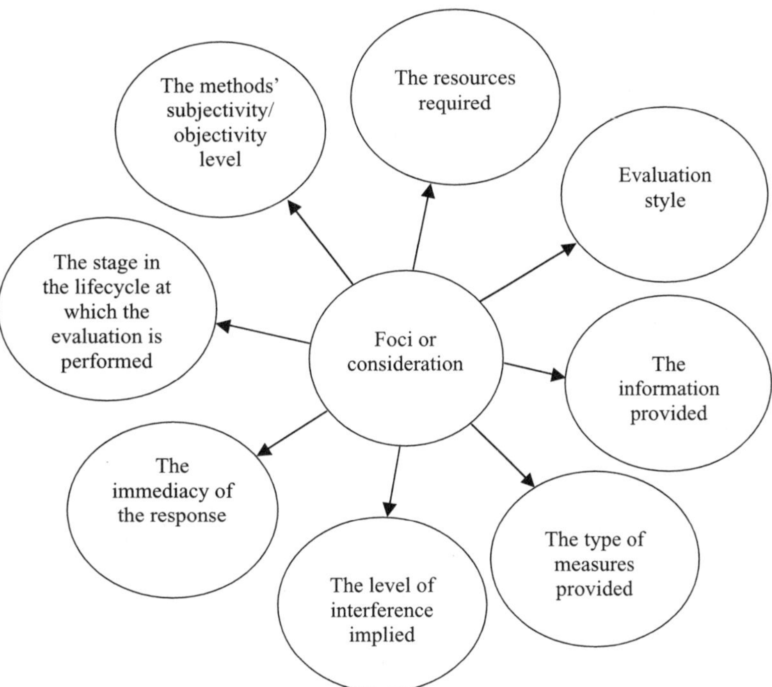


**Fig. 10.3** Foci or considerations for selecting an appropriate usability evaluation/inspection method.

The pluralistic walkthrough makes use of group meetings where usability experts, developers, and students step through a learning scenario and discuss each and every dialogue element. Furthermore, this feature inspection lists items such as sequences of features employed to perform typical tasks, difficult steps, checks for rather long sequences, and steps that require extensive experience to properly assess a proposed set of features. Some of the benefits of the pluralistic walkthrough are easy to use and learn, useful to meet the criteria of all involved parties, and allows iterative testing/evaluation [13].

Heuristic evaluation involves usability experts who determine whether each and every dialogue element satisfies set usability principles in an effective manner. Some of the advantages of heuristic evaluation are useful to identify problems early in the design process, inexpensive to implement, and easy to learn and use [13]. Its main limitation is requiring a debriefing session to find out how to fix problems. The guidelines checklist helps to ensure that the appropriate usability principles will be considered in software design work. Moreover, the checklist provides inspectors with a basis to compare the software product under consideration. Normally, checklists are used in conjunction with a usability inspection/evaluation approach.

Standards inspection involves usability specialists who inspect the interface for compliance with specified standards. These standards could be user interface standards, domain-specific software standards, or departmental standards (if any). Cognitive walkthrough makes use of a detailed procedure for simulating task execution at each and every step of the dialogue to determine if the simulated user's goals and memory content can be safely assumed to result in the next correct anticipated action. Two important advantages of the cognitive walkthrough are an effective tool for predicting problems and capturing cognitive process. Similarly, its main disadvantages are the need to train a skilled evaluator and the fact that it is focused on one attribute of usability [13].

In the selection of an appropriate usability evaluation/inspection method or combination of methods for a particular application, the selector must take into consideration the different foci of the evaluation. Most of these foci or considerations are shown in Fig. 10.3 [14].

## 5. SOFTWARE USABILITY TESTING METHODS

There are many usability testing methods that measure system performance against pre-defined criteria according to the usability attributes suggested by the usability standards and empirical metrics [6]. Usually, in

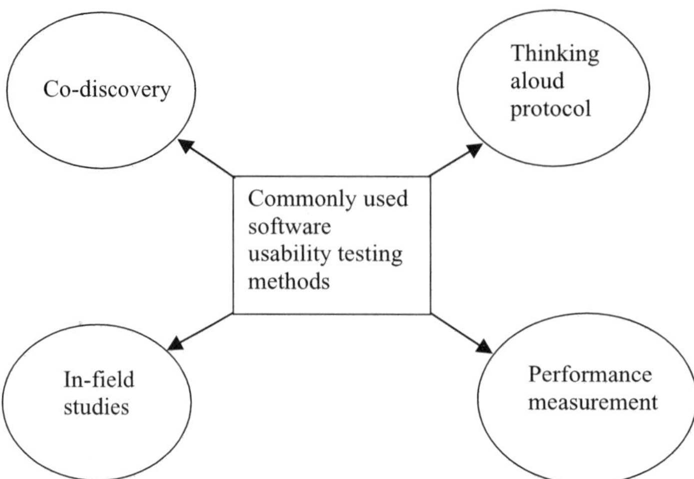


**Fig. 10.4** Common software usability testing methods

these methods, users perform specific tasks with the software product or system. Moreover, data are collected on measured performance (e.g., time required to accomplish the task). Factors such as those listed below are important with respect to these testing methods [15]:

- Selection of appropriate users.
- Selection of representative tasks to be performed by users.
- A well-designed usability laboratory.
- A well-organized usability laboratory.

Four widely used software usability testing methods are shown in Fig. 10.4 [6].

Thinking aloud protocol is a commonly used method in software usability testing because it helps to perform formative evaluation in usability tests [16]. More specifically, during the test, participants are invited to express their thoughts, opinions, and feelings while interacting with the software and performing tasks. These user remarks provide significant insight into the most suitable approach of designing the system interaction. However, thinking aloud protocol could be difficult to use with some user groups such as young students, who are distracted by the process.

The in-field studies approach is concerned with observing the users carrying out their tasks in their normal work/study environment. The main advantage of the approach is the natural user performance and group interaction. However, it has limitations in terms of measuring performance because appropriate testing equipment cannot be effectively used in typical work environments.

In the case of the co-discovery method, a group of users perform tasks together that simulate a typical work process under observation. Moreover, most of these individuals have someone else available for help. All in all, this method is considered quite useful in many work scenarios.

In the case of the performance measurement method, usability tests are directed at determining hard and quantitative data. Mostly, these data are in the form of performance metrics (e.g., required time to perform certain tasks). A usability evaluation approach based on measured performance of pre-determined usability metrics is promoted by the International Organization for Standardization (ISO) [17].

## 6. USEFUL GUIDELINES FOR PERFORMING SOFTWARE USABILITY TESTING

Over the years, professionals working in the usability area have developed various useful guidelines for conducting software usability testing. Some of these guidelines are as follows [13]:

- **Keep the testing session neutral, if possible.** It basically means that there should be no vested interest, whatsoever, in the outcomes of the test one way or the other.
- **Assist individuals participating in the usability test only in exceptional circumstances.** More specifically, let these participants struggle as much as possible.

- **Use the "thinking aloud" approach as appropriate.** This method has proven to be quite effective to capture the thinking of the participating individuals while working with the interactive software.
- **Treat each individual participating in the test as a completely new case.** More specifically, consider that each participant is unique regardless of his/her background or previous performance in usability testing sessions.
- **Use humour as appropriate to keep the test environment fairly relaxed.** Humour is quite useful to counteract participating individuals' self-consciousness, and it can help them to relax.
- **Ensure that the person conducting/directing the usability test is conscious of his/her voice and body language.** This is very important because it is quite easy to unintentionally influence someone through voice and body language.
- **Ensure that the individual conducting/directing the usability test keeps going even when he/she makes a mistake.** More specifically this person should not panic even when he/she has inadvertently revealed information or in some other way has biased the ongoing usability test. If he/she just continues, then his/her action may not even be observed by the individuals participating in the test.

## 7. PROBLEMS

1. Discuss the factors for considering usability during software development.
2. Discuss the four basic principles of the human-computer interface with respect to software.
3. Describe the software usability engineering process.
4. Discuss the steps for improving the usability of a software product.
5. Describe the following software usability inspection methods:
   - Heuristic evaluation
   - Standards inspection
6. What is the difference between pluralistic walkthrough and cognitive walkthrough?
7. List at least six important foci or considerations for selecting an appropriate usability evaluation/inspection method.
8. What are the commonly used methods in software usability testing?
9. List and discuss at least six useful guidelines for conducting software usability testing.
10. What is the difference between software evaluation and software testing?

## 8. REFERENCES


1. Keene, S. J., Software Reliability Concepts, Proceedings of the Annual Reliability and Maintainability Symposium Tutorial Notes, 1992, pp. 1–21.
2. Hopcroft, J. E., Kraft, D. B., Sizing the U.S. Industry, IEEE Spectrum, December 1987, pp. 58–62
3. Myers, B., Robson, M. B., Survey on User Interface Programming, Proceedings of the ACM CHI'92, 1992, pp. 10–15.
4. ISO/IEC 14598-1, Software Product Evaluation: General Overview, International Organization for Standardization (ISO), Geneva, Switzerland, 1999.
5. Juristo, N., Windl, H., Constantine, L., Introducing Usability, IEEE Software, January/February 2001, pp. 20–21.
6. Avouris, N. M., An Introduction to Software Usability, Report, ECE Department, University of Patras, Rio Patras, Greece, 2002.
7. Good, M., Software Usability Engineering, Digital Technical Journal, No. 6, February 1988, pp. 125–133.
8. Good, J., Whiteside, D., Wixon, D., Jones, S., Building a User-Derived Interface, Communications of the ACM, Vol. 27, October 1984, pp. 1032–1043.
9. Whiteside, J., et al., How Do People Really Use Text Editors? SIGOA Newsletter, No. 3, June 1982, pp. 29–40.
10. Wixon, D., Bramhall, M., How Operating Systems Are Used: A Comparison of VMS and UNIX, Proceedings of the Human Factors Society 29th Annual Meeting, 1985, pp. 245–249.
11. Emerson, M., Porter, D., Rudman, F., Improving Software Usability: A Manager's Guide, Report, Enervision Media, Inc., East Chatham, New York, 2002.
12. Fitzpatrick, R., Strategies for Evaluating Software Usability, Report, Department of Mathematics, Statistics and Computer Science, Dublin Institute of Technology, Dublin, Ireland, 2002.
13. Lee, S. H., Usability Testing for Developing Effective Interactive Multimedia Software: Concepts, Dimensions, and Procedures, Educational Technology and Society, Vol. 2, No. 2, 1999, pp. 100–113.

14. Dix, A., Finlay, J., Abowd, G., Beale, R., Human-Computer Interaction, Prentice Hall, Inc., Hemel Hempstead, UK, 1998.
15. Preece, J., Rogers, Y., Sharp, H., Benyon, D., Holland, S., Carey, T., Human Computer Interaction, Addison-Wesley, Inc., Reading, Massachusetts, 1994.
16. Ferre, X., Juristo, N., Windl, H., Constantine, L., Usability Basics for Software Developers, IEEE Software, January/February, 2001, pp. 22–30.
17. ISO 9241-11 (Draft International Standard), Ergonomics Requirements for Office Work with Visual Display Terminals (VDT), Part 11: Guidance on Usability, International Organization for Standardization (ISO), Geneva, Switzerland, 1997.

# Chapter 11

# WEB USABILITY

## CONTENTS



## 1. INTRODUCTION

The World Wide Web (WWW) was released just over a decade ago by the European Laboratory for Particle Physics (CERN). Its growth mushroomed to over 38 million sites in 2002 from mere 623 sites in 1993. Usage of the Web in the international economy has become an instrumental factor. For example, in 2001 the global e-commerce market was estimated to be around $1.2 trillion. Moreover, it is predicted that by 2005, there are going to be 1 billion Web users around the world. Web usability may simply be described as allowing the user to manipulate features of Website to accomplish a specific goal [1].

Today, usability rules the Web, because if a Website is not easy to use, people simply leave and move on to something else. In terms of business, it simply means that if a customer cannot find a product then he/she will not buy it. Thus, it may be said that usability has transformed from a "nice to have" into a "must have" element of e-commerce application design [2].

ISBN: 1-58883-084-5 / $35.00

With limited data on Web application user interface and usability standards, human factors or usability specialists are working toward determining ways to balance approaches such as those listed below to Web-based usability engineering [2].

- Relying on emerging Web design-associated standards.
- Performing studies on wants and needs of Web application users.
- Transferring conventional application design-based approaches to the Web environment.

This chapter presents various important aspects of Web usability.

## 2. WEB USABILITY-RELATED FACTS AND FIGURES

Some of the facts and figures directly or indirectly concerned with Web usability are as follows:

- There are over 800 million pages on the Web in the United States [3].
- In 2000, over 50% of the companies sold their products online in the United States [4].
- User interface accounts for 47% − 60% of the lines of system or application code [5].
- A study of 20 journalists who attempted to use the press areas of 10 corporate Websites to obtain basic information on areas such as the company's financial health, management, and commitment to social responsibility; revealed that the journalists were successful only 68% of the time [6].
- It is estimated that around 10% of users scroll beyond the information that is visible on the screen when a page appears or comes on [7].
- Web usability is improving somewhere between 2% and 8% each year [8].
- It is estimated that an average mid-sized company could save around $5 million annually in employee productivity by improving its Intranet (the system only used by the company employees) design [9].
- A study of e-commerce sites revealed that only 56% of intended tasks were performed successfully by the site users [10].
- A study reported that 40% of all Website visitors/users elected not to return to a site because of design-related problems [11].
- A research study revealed that around 65% of all online shopping trips end up in failure [12].
- A study reported that around 70% of retailers lacked a well-defined e-commerce strategy and believed they were using their Website to test the waters for online demand [13].

## 3. COMMON WEB DESIGN ERRORS

Over the years, professionals working in the area of Web design have identified basic errors that are common on all levels of Web design. Some of these errors are discussed below [14]:

- **Information architecture error.** This is structuring the Website to mirror the company structure instead of structuring it to mirror the users' tasks and their views of the information space.
- **Content authoring error.** This is writing in the usual linear style instead of specifically writing for online readers by keeping in mind that these readers often scan text and require rather short pages with secondary information relegated to supporting pages.
- **Business model error.** This is treating the Web as a Marcom brochure instead of a basic shift that will change conducting business transactions in the age of network economy.
- **Project management error.** This is managing a Web project as a conventional corporate project instead of managing it as a single customer-interface project. The main disadvantage of the traditional approach is that it leads to an internally focused design with an inconsistent user interface.
- **Linking strategy error.** This is treating a Website as indispensable, thus not providing appropriate links to other sites as well as not having effective entry-points for others to link to. Moreover, many companies even overlook the use of appropriate links to their own site in their own advertisements.
- **Page design error.** This is creating attractive pages for evoking positive feelings when used within the company instead of designing for an optimal user experience under a realistic environment, although the pages are less attractive.

## 4. WEB PAGE DESIGN

This is the most immediately visible element of Web design and is an important factor in the effectiveness of Web usability. Some of the key usability "dos" and "don'ts" with respect to page design are as follows [15]:

### "Dos"

- Design Web pages such that the browser can easily resize them to satisfy his/her particular requirement. For example, the available display space.
- Consider screen real estate as a very valuable commodity.
- Tailor images to elements that are meaningful. More specifically, past experiences indicate that dense graphics alienate many users.
- Aim to keep the size of most Web pages to a level that can be downloaded within 10 seconds.
- Aim to fit main page contents in the potential browser window's width, even under the circumstance when the window is not "maximized" to fill the complete screen.
- Use visual highlighting as appropriate to draw users' attention to pertinent information.

### "Don'ts"

- Avoid using animation unless it is absolutely essential.
- Avoid assuming that all potential users can see what you see.
- Avoid specifying fonts using absolute sizes.
- Avoid getting carried away with "artistic" or creative fonts.
- Avoid using all capital letters.

Some of the most important factors to be considered in page design are discussed below in detail [15].

### 4.1. Page Size

This plays an important factor in usability effectiveness. More specifically, it is important to usability in basically two ways: the flexibility of the pages to fit the available display area, and the downloading and displaying speed of pages. Both of these are discussed below, separately [14, 15].

### Page Flexibility

As this is a key factor in usability effectiveness, the following guidelines can be quite useful in its successful achievement:

- Design Web pages in such a manner so they can easily be resized (i.e., to fit within a fairly wide range of window sizes).
- Make use of relative, instead of absolute, sizes for elements that fall under the resizing capability of the browser.
- Pay close attention to resize-ability when designing footers, headers, etc.
- Usually, no horizontal scrolling is necessary if the window is 800 pixels wide.
- Ensure that all the important page elements are visible with scrolling when the window is 400 pixels high.

### Page Downloading Speed

The length of time to download a page from the server and display it in the browser window is a key factor to sizing Web pages effectively. Nonetheless, response time is defined as the time from when a user requests a page to when it has been displayed completely. Some of the useful guidelines associated with response time are as follows:

- Ensure that the response time is within 0.1 second to make system "feel interactive".
- Ensure that the response time is within 10 seconds to keep user's attention.
- Ensure that the response time is within 1 second to fit into the user's chain of thoughts.

- Provide an adequate warning to users when a Web page will require greater than 10 seconds to download.

All in all, past experiences indicate that fast Websites usually get more traffic and have lower bailout rates.

### 4.2. Font Usage

Fonts are used to create various types of Web page elements, including headers and footers, menus, navigation bars, tables, links, and buttons, in addition to the text that conveys most of Website's contents. Font faces fall under two basic categories: serif and sans-serif [15]. Serif fonts have small appendages at the tops and bottoms of letters. Some examples of these fonts are Courier, Times Roman, and Century. These fonts are helpful to make it easier to read long lines. Sans-serif fonts are simpler in shape because they consist of only primary line strokes. Typical examples of these fonts are Arial, Helvetica, and Futura. Often sans-serif fonts are used for short phrases such as button labels, titles, and outlines. Some of the directly or indirectly font usage-related pointers are as follows:

- Avoid specifying absolute font sizes.
- Different browsers support different font faces.
- Use Italics to define terms or emphasize an occasional word.
- Avoid getting carried away in the use of font faces, sizes, and styles.

### 4.3. Image Usage

The creativity with which images of all types (e.g., plots, diagrams, and photographs) can be integrated with textual elements is considered to be an important factor for the popularity of the Web. On the other hand, users usually blame images for many of the usability-related problems that plague Web access. Nonetheless, some useful guidelines for using images are as follows [14, 15]:

- Use animation in only those situations in which it truly adds to the meaning of the information.
- Use the most efficient format for an image under consideration.
- Try to reduce the image resolution as much as possible.
- Try to limit graphics to elements that are really required.
- When you use a graphic, try every effort to re-use it on other pages if the need arises.
- Avoid using a different photograph on each page of a Website. Otherwise, the Website performance will be sacrificed.
- Limit the use of the number of different colors.
- Aim to include a thumbnail image on the Web page that links to the larger one.
- Try to use a commercial image compression tool to reduce the size of image files as much as possible.

### 4.4. Textual Element Usage

Most of the core elements of a Website are conveyed through the use of text, tables, and lists. Therefore, it is very important to write in a style that not only transmits desired information, but also reflects how Websites are actually utilized. Some of the useful guidelines for writing effective Web page text are as follows [15]:

- Ensure that the text is as concise as possible.
- Ensure that subjective or exaggerated language is converted to more neutral terms.
- Ensure that the text layout is converted to a format that is more "scan-able".
- Combine the three previous approaches or factors to obtain maximum benefits.

### 4.5. Help for Users

Usually, Web users do not read Web pages in a serial manner. They hop from one visual element to another. Thus, the biggest challenge faced by Web page designers is using the visual elements effectively in order to draw users' attention to key contents. More specifically, it may be said that potential users would not read text unless they are enticed to do so. In this regard, some of the key guidelines/pointers are as follows [15]:

- Reinforce the hierarchy of Web page contents with a hierarchy of visual dominance.
- Make use of size to help users understand which elements fall where with respect to the content hierarchy.
- Make use of white space to create distance between groupings as well as to group related elements together.
- Ensure that the most important text has the greatest possible contrast.
- Make use of the same color in developing a common thread among elements that cannot be placed next to each other.
- Past studies indicate that items above and to the left of the page center appear to be noticed first.
- Past studies indicate that users normally assume that a row of similar elements should be "read" from top to bottom or from left to right.
- Ensure that the use of visual highlighting methods is consistent across the entire Website.
- Test each Web page's final design by eliminating all visual elements in question, one at a time.

## 5. WEBSITE DESIGN

Usually greater attention is given to page design than site design. However, from the usability perspective, the site design is more challenging and important. Some of the key usability "dos" and "don'ts" in regard to Website design are as follows [15]:

### "Dos"

- Ensure that pages of a single Website share a common look-and-feel as much as possible.
- Ensure Web page support browser resizing as much as possible.
- Ensure that all the Web pages honour the user's browser settings.
- Ensure that each Web page includes real content.

### "Don'ts"

- Avoid using frames.
- Avoid having a "banner page".
- Avoid saying "welcome" on Web pages.
- Avoid having a copyright notice. It is not required to establish ownership to Web materials.
- Try not to pop up windows without the user's consent.

Some of the key factors to be considered in Website design are discussed below in detail [14, 15].

### 5.1. Site Organization

Site organization requires careful consideration during the design process because users do not read Web pages the way they read books. Some of the guidelines directly or indirectly concerned with Website organization are as follows [14, 15]:

- Provide users at least some content on each and every page.
- Ensure that all pointers to related topics are visible somewhere in the upper-half of the page under consideration.
- Organize the site into a number of bite-size pieces capable of being traversed in different ways to take advantage of the Web's navigational flexibility.
- Ensure that the important information is positioned in such a manner so that it is clearly visible even when the browser window is shrunk to about 50% of the screen width.
- Avoid requesting potential users to provide email addresses.
- Avoid displaying blocks of text in a large font.

### 5.2. Site Pages' Shared Elements

For the effective usability of a site, it is important to help users become familiar with the site with minimum effort on their part. This can be accomplished by adopting a consistent page style that repeats common elements throughout the Website. This process can also help to improve user speed. For example, re-using the same image on each and every page improves speed to a certain degree. Another factor that helps usability is to concentrate all common elements at the top and bottom of each page or along the left-hand side.

According to some usability studies, some user expectations of common elements are as follows [14, 15]:

- Ability to go to the home page by simply clicking on the Website's icon.
- Clear displaying of a "contact us" mechanism.
- Gathering of information intended for sponsoring agencies under "about us".
- Inclusion of a help feature for only those circumstances in which it provides substantive information.
- Inclusion of a search mechanism only if the site includes a very large number of pages.

### 5.3. Site Testing and Maintenance

In order to maintain usability effectiveness, regular testing and maintenance of the Website are necessary. Moreover, this helps to ensure that the users see the intended message.

To ensure Web page quality, test the design and each new page by using at least Internet Explorer and Netscape, disabling images, using different browser window widths, and using dialup connection [15].

As the Web pages tend to change over time, it is absolutely essential to perform maintenance activities on a regular basis. At least once a month, the verification of all links still being active must be performed. Moreover, whenever a Web page is modified, every effort should be made to double check that its links are functioning normally.

## 6. NAVIGATION AIDS

These are used by users to find their way around Websites. Some examples of these aids are links, menus, navigation bars, and buttons. Over the years, various navigation aid usability-related studies have been conducted. Some of the key "dos" and "don'ts" related to the usability of these aids are as follows [15, 16]:

#### "Dos"

- Organize navigation aids by carefully considering the user tasks to be performed.
- Give utmost importance to navigation bars or bread-crumb trails because they provide users with an understanding of location on a site.
- Ensure the conformance to the standard practice of having all links underlined.
- Ensure that all links, menu items, or navigation bar items that lead to the "current location" are deactivated properly.

#### "Don'ts"

- Avoid labelling menu items, buttons, or links with meaningless phrases.
- Avoid assuming that all potential users will be able to familiarize themselves with the Website.
- Avoid changing the standard colors used for links.
- Avoid confusing users by implementing menus in a "creative" fashion.

Some of the important factors to be considered with respect to the navigation aids are described below [15, 16].

### 6.1. Link Usage

Links are probably the most common mechanism supporting Website navigation. Web pages use links in the following three ways [15, 16]:

- To direct users to an alternative source, in the event the current page does not contain the required information.

- To provide efficient access to other pages of the Website.
- To direct users to pages containing additional information about the graphic/text mentioned in the link.

Some of the important guidelines associated with the effective use of links are as follows [15, 16]:

- Underline all links and use the standard colors.
- Format links by using upper- and lower-case letters.
- Select link text with care.
- Underline the words that really matter to improve link readability.
- Group the links into categories when multiple links appear in a list.
- Avoid underlining the non-link text.
- Locate all alternative links at the top of the page.
- Make the image itself be the link if there is a need to link to a larger copy of an image.

### 6.2. Navigation Bar Usage

The main purpose of having navigation bars is to lay out the Website structure in a hierarchical form. They are usually located along one side of the Web page. Sometime, they may also be found in a box at the top of the Web page's main content area. There are two key factors that have proven very useful in using the navigation bar effectively: (i) The identification of the top 10 things, during the navigation bar development, that potential users are likely to do on the site under consideration; and, (ii) The selection of navigation structure and labels with care.

The following steps are considered quite useful for selecting navigation structure and labels [15, 16]:

- Develop a list of all the operations/functionalities the Website has to support for users to accomplish the identified top 10 things/tasks effectively.
- Write down the operations on individual index cards.
- Spread out the cards and group them into logical classifications.
- Identify at least five potential users and have them to repeat the three previous steps.
- Make a comparison of the results of all the sorting. If some pattern of classification is not emerging, repeat the entire process (i.e., all the steps) again.
- Once the classifications arrived at by most potential users appear to be similar, take advantage of them to develop an outline of the Website structure.
- Present the structure outline to at least five other potential users and ask for their input. Repeat the process as required.

### 6.3. Menus and Menu Bar Usage

Usually, menu bars are used by Websites to provide basic navigation functionality. They may include various links or contain menu titles that drop down as users' cursors pass over them or click on them. Some of the important guidelines concerning menus/menu bars are as follows [15, 16]:

- Ensure that menu titles are short and form a consistent group.
- Anchor all menus to a menu bar across the top of the Web page.
- Ensure that the menu is obviously a menu by making the title look like a typical link.
- Format menu titles and menu items by using upper- and lower-case letters.
- Group menu items together logically.
- Avoid using cascading (i.e., multi-level) menus.

## 7. USABILITY TOOLS

The success of a Website very much depends on good usability. The evaluation of Web usability using conventional usability engineering methods could be quite difficult because today's users are heterogeneous and geographically dispersed. There are a number of modern approaches or tools that can highlight potential usability-related problems. More specifically, these tools are good for checking routine site-design elements

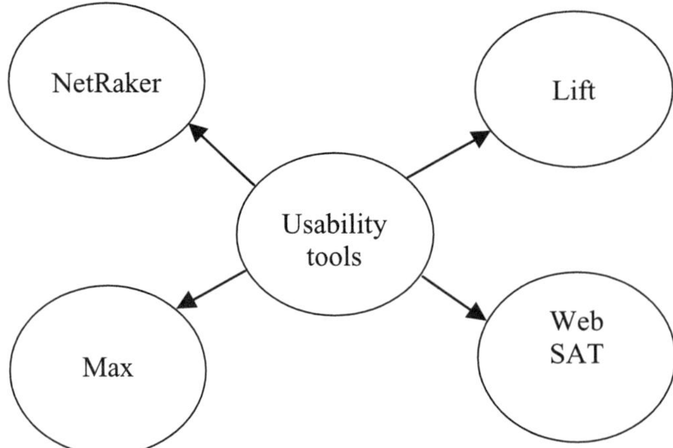


**Fig. 11.1** Modern usability tools

with respect to consistency, as well as for encouraging the application of good design practices. Four of the tools are shown in Fig. 11.1 [17]. Each of these tools is described in detail below.

### 7.1. Web SAT

The Web Static Analyzer Tool (Web SAT) is one of tools that belongs to the Web Metrics Suite developed by the National Institute of Standards and Technology (NIST), and it is used to check Web page Hyper Text Mark-up Language (HTML) against typical usability guidelines for potential problems. Web SAT allows up to five individual Uniform Resource Locators (URLs) to be checked against its usability guidelines. At the end of the evaluation process, Web SAT provides a report of problems uncovered on each and every Web page entered. The problems are grouped under six categories as shown in Fig. 11.2 [17]. In the accessibility category, the problems are associated with the page making proper use of tags for visually impaired users.

In the form use category, the problems are associated with the form Submit and Reset buttons. In the readability category, the problems pertain to content readability. In the maintainability category, the problems are concerned with tags and coding information that would make the page easier to port to another server. In the performance category, the problems are associated with the size and coding of graphics in relation to page download speeds. In the navigation category, the problems are associated with the coding of links. The main limitation of the Web SAT is that is can check only individual Web pages.

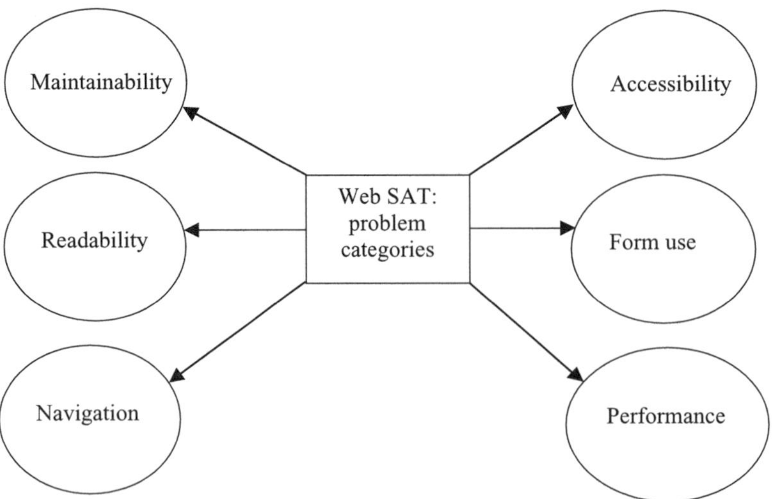


**Fig. 11.2** Web SAT: usability-related problem categories

### 7.2. NetRaker

This is composed of various online tools that help identify usability-related problems and conduct market research. NetRaker provides comprehensive guidelines for composing objective survey questions and a customizable set of usability survey templates. The questions are randomly made available to Website users by giving them an option to participate. The survey requires users to carry out tasks on the Website and then provide adequate feedback with respect to the easiness of performing the tasks. Some of the advantages of NetRaker are as follows [17]:

- Useful in obtaining feedback in the context of Website's intended purpose as opposed to relying on generic HTML checks or statistical analysis.
- A useful tool to survey users and gather usability feedback quickly.
- The automation of NetRaker ensures that users are surveyed consistently.

All in all, NetRaker is one of the best tools to highlight usability-related issues because it is based on direct feedback of users.

### 7.3. Max

Max is another useful usability tool that scans through a Website to collect information for vital statistics and rates a site's usability. Max makes use of a statistical model to simulate a user's experience in calculating ratings in the following three areas [17]:

- Load time
- Accessibility
- Content

In the case of load time, Max estimates the average time taken to load Website pages. In the case of accessibility, Max estimates the average time a user takes to find something on the site. Finally, in the case of content, Max summarizes the percentage of different media elements (i.e., multimedia, text, and graphics), as well as client-side technologies employed (e.g., Flash and Portable Document Format (PDF)) that comprise the Website under consideration. The main strength and weakness of the Max are that it provides a performance benchmark and it does not provide many suggestions to make design changes, respectively.

### 7.4. Lift

This is another useful usability tool that performs analysis of a Web page for potential usability problems. There are two kinds of Lift: Lift Online and Lift Onsite. Lift Online performs HTML checks derived from usability principles in a similar manner to Web SAT. More specifically, it examines one page at a time, and then provides a report on the page's usability-related issues. Furthermore, it may be said that Lift Online goes one step further than Web SAT because it provides specific code change recommendations.

Lift Onsite can be run from a Personal Computer (PC) and it provides the very compelling feature of directly fixing the HTML problems as they are being reviewed in the usability evaluation report.

All in all, it may be said that Lift provides usability-based HTML validations for insuring good coding practices.

## 8. GENERAL QUESTIONS FOR EVALUATING THE EFFECTIVENESS OF THE WEBSITE MESSAGE COMMUNICATION

This section presents a checklist of some general questions for evaluating the effectiveness of communicating the Website message. These questions are grouped under six distinct areas as shown in Fig. 11.3 [18–21]. Questions belonging to each of these areas are presented below. All in all, these questions are considered to be useful in improving Web usability directly or indirectly.

### Content

- What type of information is the Website expected to convey?
- Does the site offer any communication options?

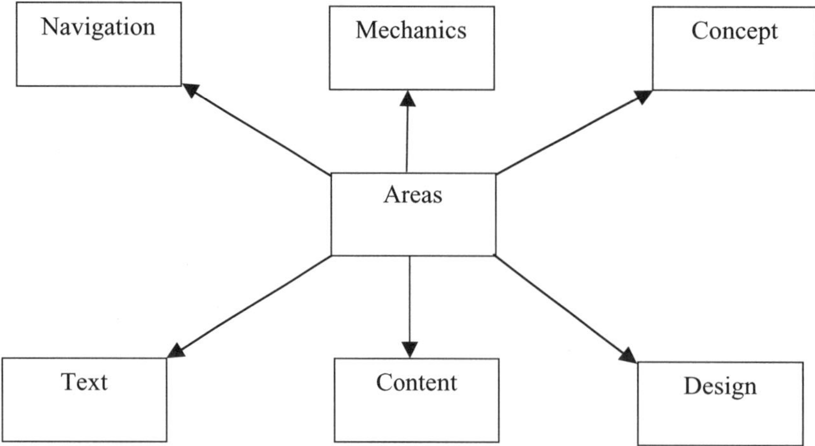


**Fig. 11.3** Areas of grouping questions for determining website message communication effectiveness

- How unique and accurate are the contents of the Website?
- Is the information on the Website accurate, clear, complete, attractive, and accessible?
- What portion of the Web page(s) was allocated for content compared to other factors?
- Is the content easily accessible and the purchasing procedure (if applicable) user friendly?

### Text

- Are the button and hyperlink titles clear?
- Is the text grammatically checked?
- Can the text be read in a cursory manner effectively?
- Does the first page clearly convey a message to visitors, including what they can expect to find in the Website?
- Is the text on first and feature pages brief enough for its effective use?
- Are the titles appropriate for use by search engines effectively?
- Are the subheadings and titles informative for their effective use?
- Is the text sufficiently attractive to read?

### Design

- Is the site design impressionable?
- Can the design in any way deter visitors?
- Is there an appropriate balance between design and content?
- Are the visitors directed in an effective manner to the most important page elements?
- Is the site style unique?
- Is there sufficient contrast between text color and background?

### Concept

- What image of the organization does the site project?
- What expectations will the Website raise for the potential visitors?
- What does the first page promise concerning the rest of the Website?
- Can the rest of the site meet this promise effectively?
- What existing Websites can be compared with the one under consideration?

### Navigation

- Is it possible to predict the contents of the options on the menu without clicking them?
- How does the site conform to current Web standards?
- Does the site have a local search engine?

- Is it possible to find anything using simple and straightforward keywords on the engine?
- How easy is it to use the menu, buttons, and hyperlinks to browse the site effectively?
- Is it possible that visitors will ever be bewildered by the appearance of something unexpectedly?
- Will it always be clear to visitors about their browsing location?

### Mechanics

- Are all the hyperlinks and buttons functioning as per requirements?
- How fast does the site react and the pages load?
- How functional is the Website?
- Are there any error messages?
- Do tools such as roll-down menus and mouse-over (if used) support the use of the site?

## 9. PROBLEMS

1. Define the term "Web usability".
2. List at least six facts and figures concerned with Web usability.
3. Discuss the following common Web design errors:
   - Content authoring error
   - Linking strategy error
   - Information architecture error
4. Discuss the key "Dos" and "Don'ts" associated with Web page design.
5. Discuss the following terms:
   - Page downloading speed
   - Page flexibility
   - Font usage
6. List at least six key "Dos" and "Don'ts" associated with Website design.
7. Discuss the key "Dos" and "Don'ts" associated with navigation aids.
8. Describe the following:
   - Link usage
   - Navigation bar usage
   - Menus and menu bar usage
9. Describe the following two usability tools:
   - Web SAT
   - NetRaker
10. Write an essay on Web usability.

## 10. REFERENCES

1. Powell, T., Web Design: the Complete Reference, Osborne McGraw-Hill, Berkeley, California, 2000.
2. Cloyd, M. H., Designing User-Centered Web Applications in Web Time, IEEE Software, January/February 2001, pp. 62–69.
3. Preece, J., Online Communities: Design Usability, Supporting Sociability, John Wiley and Sons, New York, 2000.
4. McDonald, S., Waern, Y., Cockton, G., editors, People and Computers XIV: Usability or Else!, Springer-Verlag, London, 2000.
5. Trenner, L., Bawa, J., Editors, The Politics of Usability: A Practical Guide to Designing Usable Systems in Industry, Springer-Verlag, London, 1998.
6. Nielsen, J., Corporate Websites Get a 'D' in PR, Alertbox, April 1, 2001. Available online at www.useit.com/alertbox/20010401.html.
7. Nielsen, J., Top Ten Mistakes in Web Design, Alertbox, May 1996. Available online at www.useit.com/alertbox/9605.html.
8. Nielsen, J., PR on Websites: Increasing Usability, Alertbox, March 10, 2003. Available online at www.useit.com/alertbox/20030310.html.
9. Nielsen, J., Intranet Usability: The Trillion-Dollar Question, Alertbox, November 11, 2002. Available online at www.useit.com/alertbox/20021111.html.

10. Chi, E. H., Improving Web Usability Through Visualization, IEEE Internet Computing, March/April 2002, pp. 64–71.
11. Manning, H., McCarthy, J. C., Souza, R. K., Why Most Web Sites Fail, White paper, Forrester Research, Cambridge, Massachusetts, September 1998.
12. Souza, R. K., et al., The Best of Retail Site Design, White paper, Forrester Research, Cambridge, Massachusetts, October 2000.
13. Becker, S. A., Mottay, F. E., A Global Perspective on Web Site Usability, IEEE Software, January/February 2001, pp. 54–61.
14. Nielsen, J., Designing Web Usability: The Practice of Simplicity, New Riders Publishing, Indianapolis, Indiana, USA, 2000.
15. Brown, G. E., Web Usability Guide, Report, NEES Consortium, Inc., Richmond, California, 2003. Available online at www.nees.org/info/contact_us.html.
16. Fleming, J., Web Navigation: Designing the User Experience, O'Reilly and Associates, Inc., Sebastopol, California, 1998.
17. Chak, A, Usability Tools: A useful Start, New Architect: Strategy Product Review, August 2000. Available online at www.webtechniques.com/archives/2000/08/stratevu/.
18. Wilson, J., Is Your Site Effective?, Communication: Web Site, Columnar.com, Los Angeles, 2000. Available online at www.columnar.com/communication.html.
19. Niederst, J., Web Design in a Nutshell: A Desktop Quick Reference, O'Reilly and Associates, Inc., Sebastopol, California, 2001.
20. Williams, R., Tollett, J., The Non-Designer's Web Book: An Easy Guide to Creating, Designing and Posting Your Own Web Site, Peachpit Press, Berkeley, California, 2000.
21. Price, J., Price, L., Hot Text: Web Writing that Works, New Riders Publishing, Indianapolis, Indiana, USA, 2002.

# Chapter 12

# MEDICAL DEVICE USABILITY

## CONTENTS



## 1. INTRODUCTION

Each year, a vast sum of money is spent on healthcare throughout the world. For example, in the Untied States alone in 1994, medical spending was in the order of $938 billion [1, 2]. That means that $1 out of every $7 spent that year in the United States was on healthcare. Today, there are around 6,500 hospitals and 700,000 doctors in the U.S., and U.S. exporters control half of the world's $71 billion medical device market [1, 2].

Good user-interface design of medical device/equipment is critical for effective and safe device/equipment installation, operation, and maintenance [3, 4]. In fact, the results of various studies performed over the years indicate that poorly designed human-machine interfaces of medical device/equipment significantly increase the risk for the occurrence of human errors [3, 5, 6]. These errors can lead to patient injury or even death.

It may be added that medical device/equipment usability concerns, with respect to potential users such as patients, physicians, nurses, family members, and professional caregivers, must be raised to the same level as traditional/conventional technological, manufacturing, and economic concerns, during the design phase. This chapter presents various important aspects of medical device usability.

## 2. MEDICAL DEVICE USERS, USE ENVIRONMENTS, USER INTERFACES, AND USE DESCRIPTION

A medical device may be easy for one group of users to use but difficult for another user group. Users need devices that can be used effectively and safely. In order to satisfy these user needs, it is necessary to fully understand the abilities and limitations of all types of intended users. Some examples of medical device users are young and old individuals, professional healthcare providers, and patients. Factors such as stress, fatigue, and medication can affect the ability levels of device users.

ISBN: 1-58883-084-5 / $35.00

Nonetheless, some of the important characteristics of the user population that should be taken into consideration during medical device design are as follows [7]:

- General health and mental state (e.g., stressed, tired, and relaxed) when using the device under consideration.
- Memory, strength, physical size, and cognitive ability.
- Past experience with similar devices or user interfaces.
- Sensory capabilities (i.e., touch, vision, and hearing).
- Ability to adopt to adverse circumstances.
- Coordination (i.e., manual dexterity) and motivation.
- Knowledge about device operation and the related medical condition.

The use environments of medical devices can vary quite significantly and can have major impacts on their usability. Factors such as those listed below must be considered carefully with respect to users [1].

- **Mental workload.** This is the degree of concentration and thinking an individual exerts while using a device. When the mental workload imposed on users by the environment exceeds their abilities, the probability of ineffective use of devices increases dramatically. An example of this situation could be an operating room having too many alarms on different devices, thus making it difficult for an anaesthetist to identify the source of any single alarm.
- **Physical workload.** This is associated with the device use and adds to the user stress. Under high stress, users are distracted and have less time to make decisions, consider multiple device outputs, etc.
- **Light and noise.** The effectiveness of visual and auditory displays (e.g., lighted indicators, auditory alarms, and other signals) can be limited by the use environments if they are designed poorly. For example, in noisy environments, users may not hear alarms if they are not sufficiently loud or distinctive. Nonetheless, some of the important considerations for displays, i.e., including visual alarm indicators, are viewing angles, presence of other devices in the use environment, and ambient light levels.
- **Motion and vibration.** They can affect user ability to read displayed information, to perform fine physical manipulations such as typing on the keyboard portion of a medical device, etc.

User interfaces of well-designed medical devices will facilitate correct actions and prevent or discourage the occurrence of hazardous actions. The user interface comprises all elements of a medical device with which users interact while using the device, preparing it for use (e.g., set-up and calibration), or carrying out maintenance activities. More specifically, the user interface includes all hardware features that control the operation of the device. For example, knobs, switches, buttons, and user information providing device features such as displays, indicator lights, and auditory and visual alarms.

In modern medical devices, usually the user interfaces are computer-based. In this case, interface characteristics include items such as those listed below [7].

- Control and monitoring screens
- Keyboards
- Help functions
- The manner in which data are organized and presented
- Mouse
- Prompts
- Navigation logic
- Alerting mechanisms
- Screen elements
- Data entry requirements

All in all, items such as device labelling, training materials, operating instructions, and packaging are also considered part of the user interface, and thus require a careful consideration with respect to their effective usability.

In order to understand a medical device's use accurately and completely, the description of its intended use is essential. This description includes information on areas such as [7]:

- Use environments
- Device operation

- User interface design or preliminary design
- General use scenarios, more specifically, scenarios describing how the device will actually be used
- User population characteristics, more specifically, those characteristics that can affect device use
- User needs for effective and safe use of the device and how the device satisfy them

## 3. MEDICAL DEVICES WITH HIGH INCIDENCE OF USER/HUMAN ERROR AND A GENERAL APPROACH FOR DEVELOPING EFFECTIVE USER INTERFACES OF MEDICAL DEVICES

In 1991, the Food and Drug Administration (FDA) performed a study of data collected over many years and identified the 20 most user/human error-prone medical devices [8, 9]. These devices (in the order of highest error-prone to lowest error-prone) were Glucose meter, Ballon catheter, orthodontic bracket aligner, administration kit for peritoneal dialysis, permanent pacemaker electrode, implantable spinal cord stimulator, intravascular catheter, infusion pump, urological catheter, electrosurgical cutting and coagulation device, non-powered suction apparatus, mechanical/hydraulic impotance device, implantable pacemaker, peritoneal dialysate delivery system, catheter introducer, catheter guide wire, transluminal coronary angioplasty catheter, low-energy defibrillator (external), continuous ventilator (respirator), and contact lens cleaning and disinfecting solutions.

All in all, the FDA study indicates that errors in using medical devices cause an average of at least three deaths or serious injuries each day in the United States [8].

The steps of a general approach for developing effective user interfaces of medical devices are presented in Fig. 12.1 [3, 10]. Step 2 is also concerned with allocating tasks between the humans and the system. In step 3, design prototypes and usability goals are also developed. In step 4, the results of usability testing are also evaluated against performance goals and objectives, and a loop back to step 3 is made as appropriate. In step 6, a loop back to step 3 is made as appropriate.

## 4. GUIDELINES FOR MAKING MEDICAL DEVICE INTERFACES MORE USER-FRIENDLY

Medical devices such as infusion pumps, patient monitors, kidney dialysis machines, ventilators, and blood chemistry analyzers often have various superficial user-interface design problems. These problems can

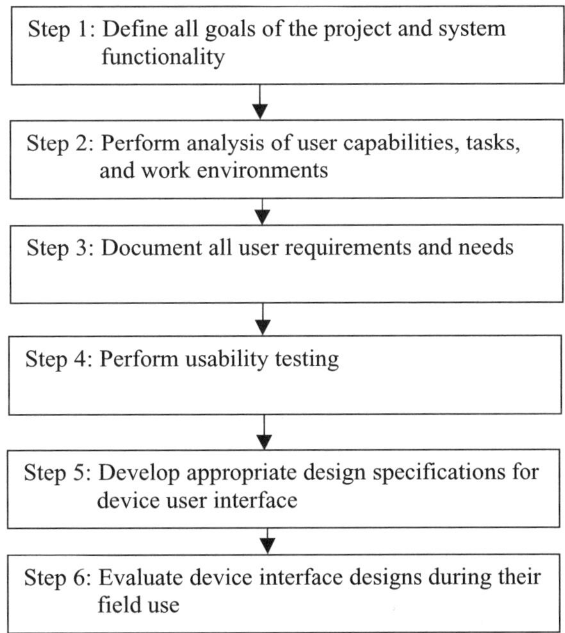


**Fig. 12.1** General approach steps for developing effective user interfaces of medical devices

negatively affect a device's usability and appeal. Past experiences indicate that such problems are relatively easy to remedy. The following guidelines are useful to address these design problems [11]:

- **Reduce screen density as much as possible.** This is concerned with reducing the over-stuffing of medical device displays with information and controls. The empty space created by such reduction is extremely useful in a user-interface because it helps to separate information into related groups and provides a resting place for eyes of users. Otherwise, overly dense-looking device user-interfaces can be quite intimidating to medical professionals such as physicians, nurses, and technicians, making it difficult for these individuals to retrieve desired information at a glance. Nonetheless, actions such as those listed below can help to eliminate extraneous information on medical device displays.
  - Use simplified graphics
  - Decrease the size of graphics related to identity (i.e., logos and brand names)
  - Present secondary information on demand through pop-ups or relocate it to other screens, if possible
  - Utilize empty space rather than lines to separate content
  - Reduce text size by stating things in a more simplified manner
- **Limit the usage of colors.** This is concerned with limiting the color palette of a device user interface. Two useful guidelines in this regard are as follows:
  - Keep the colors of the background and major on-screen components to between three and five, including shades of grey.
  - Ensure that the selection of colors is consistent with medical conventions. For example, red is widely utilized to symbolize alarm related-information or to communicate arterial blood pressure values.
- **Ascribe to a grid.** Past experiences indicate that most screens generally function and look better when their screen elements are aligned and they serve a utilitarian objective. Pay-offs in terms of visual appeal and perceived simplicity are accrued by fitting on-screen elements into a grid. Moreover, grid-based screens are generally easier to implement in computer code because of the predictability of visual elements. The following two guidelines are useful with respect to ascribing to a grid:
  - Start by defining the screen's dominant components and approximate space requirements when developing a grid structure.
  - Try to keep on-screen elements at fixed distance from the grid lines. This can be quite useful to provide visual appeal.
- **Simplify typography as much as possible.** Efficient medical device user interfaces are based on typographical rules that make screen contents easy and straight forward to read and direct end-users effectively toward the important information first. Usually, this is achieved by having a single font and a few character sizes.

  Another approached used to simplify typography is by reducing or eliminating excessive highlighting such as bolding, italicizing, and underlining. All in all, normally a single approach such as bolding or italicizing is quite sufficient to highlight important information effectively when used in conjunction with extra line spaces and different sizes of font.
- **Create effective visual balance.** Usually, this is created about the vertical axis by arranging visual elements on either side of an assumed axis. More specifically, each side contains roughly the same amount of contents as empty space. There are various methods that can be used to evaluate the balance of a composition. In turn, perceived imbalances can be corrected through a number of means, including relocating elements to other screens, reorganizing information, popping up elements only upon request, or adjusting the gaps between labels and fields.
- **Eliminate design inconsistencies as much as possible.** These inconsistencies are quite toxic to usability and user-interface appeal. Furthermore, for some medical devices, they may also compromise safety. The occurrence of these design inconsistencies can be prevented by developing and maintaining a style guide. Subsequently, this style document can be integrated into appropriate final design specifications.
- **Make use of simple language as much as possible.** Often medical device user-interfaces are characterized by overly complex words and phrases that create various types of usability problems. These can easily be reduced or eliminated altogether by the use of simple language. In this regard, some of the specific corrective measures are writing shorter sentences, using consistent syntax, giving meaningful headings and subheadings, and breaking rather cumbersome procedures into a number of ordered steps.
- **Provide appropriate navigation cues and options.** A navigator can sometimes become lost when going from one place to another in a medical device user-interface. It is much like the case when some-

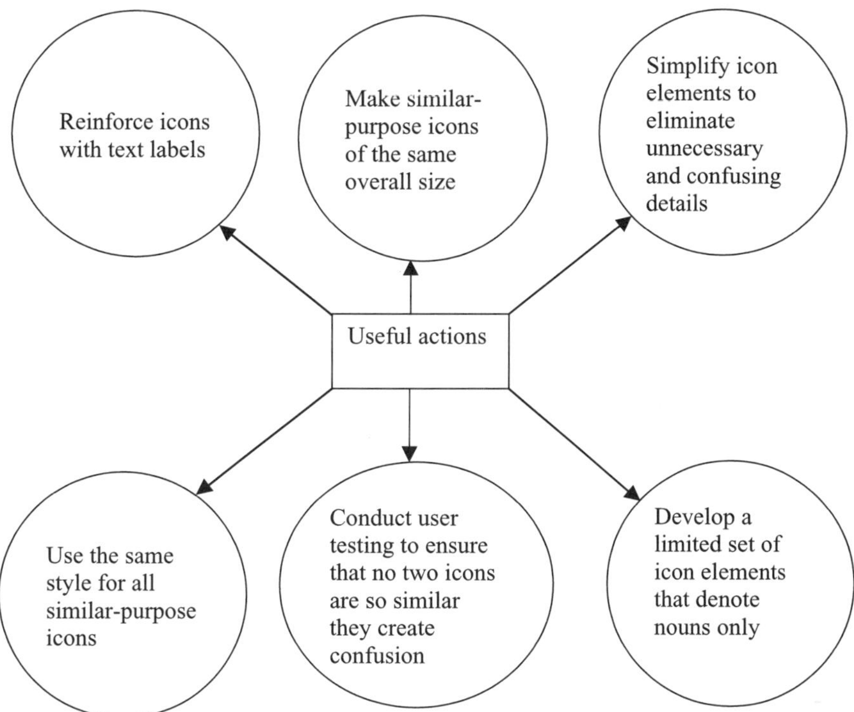


**Fig. 12.2** Useful actions for harmonizing icons and maximizing their comprehension

one is travelling in an unfamiliar city with no road signs or signs in another language than the traveller understands. More specifically, the problem results from device users not knowing where they are in the user-interface structure.

Actions such as those listed below are quite useful with respect to providing navigation cues and options.

- Place effective titles on each and every screen and subcomponent by means of a header (i.e., a contrasting horizontal bar that incorporates text).
- Number electronic documents' pages.
- Group together navigation options and controls such as "Go back", "Help", "Go to main menu", and "Next" in a single and consistent location. This will allow users to return easily to a previous screen or undo an action without any fear of getting lost.

• **Make use of hierarchical labels as much as possible.** The use of redundant labels leads to congested screens that usually take a long time to scan. Moreover, excessive visual congestion usually makes it rather cumbersome for device users to pick out even the most salient details. Hierarchical labelling is quite useful to save space and speed scanning by displaying items such as arterial blood pressure, heart rate, and respiratory rate in a more efficient manner.

• **Harmonize and refine icons.** This is a very important step. Some of the actions that can be taken to give the icons a family resemblance to each other and maximize icon comprehension are expressed in Fig. 12.2 [11].

## 5. DESIGNING MEDICAL DEVICES FOR OLD USERS AND CUMULATIVE TRAUMA DISORDER IMPLICATIONS IN MEDICAL DEVICE DESIGN

The population of older people in the United States is growing at a significant rate. For example, for the period from 1990–2000, the number of persons age 55 years and over was forecasted to increase around 11.5% (i.e., a gain of almost 5 million people) [8, 12]. Moreover, it is forecasted that by the year 2020, one out of every five or six Americans will be over 65 years of age. It means there is a definite need to design medical devices for use by older people. This requires careful consideration of factors such as those shown in Fig. 12.3 during the device design [8, 13–15].

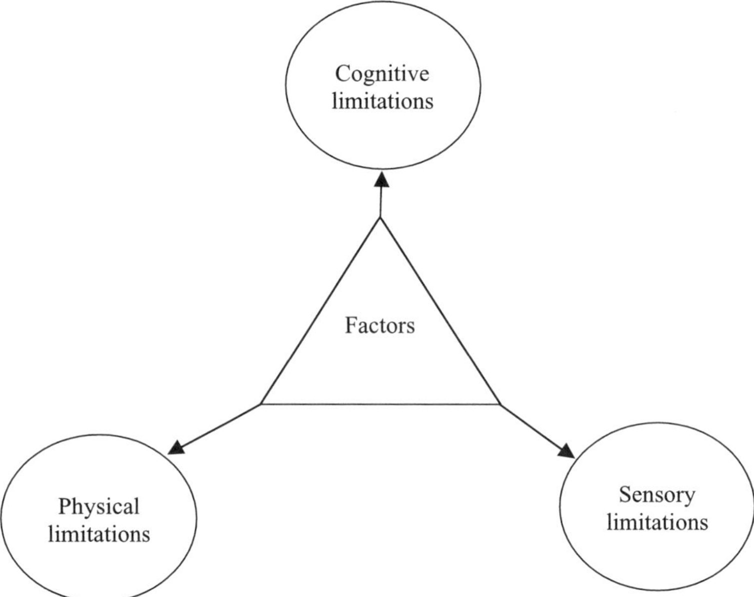


**Fig. 12.3** Factors to be considered in designing medical devices for use by older people

Old people are subject to various limitations. The two common ones are impaired vision and hearing. Some old individuals not only lose their ability to see well in low light, but also their abilities to differentiate light intensities and colors. This means that the designers must give careful consideration to using somewhat oversized fonts for displays, readouts, and labels on devices being designed for older users. The decline in a person's hearing ability is generally a function of age. Furthermore, an individual's hearing tends to decrease with the increase in sound frequency, and men and women, as they age, suffer greater hearing loss at frequencies in the 3,000–6,000 Hz range and 550–1,000 Hz range, respectively. Therefore, designers must take into consideration factors such as these in medical devices to be used by older people.

Older medical device users are also subject to various types of physical limitations. For example, a significant number of people usually lose 10% to 20% of their strength by the time they reach 60 to 70 years of age. Moreover, their mobility may also be handicapped to a certain degree by various types of joint problems and associated pain. In order to overcome physical limitations such as these, the designers should incorporate controls having large-diameter knobs into medical devices so that rotation requires a lesser degree of fine motor control, as well as textured knob surfaces requiring less pinching strength to overcome finger slippage, and so on.

Past experiences indicate that the cognition abilities of older persons may vary quite considerably. More specifically, many of these individuals may remain sharp, but others may experience attention deficits often known as "cognitive rigidity", a condition that makes it difficult to learn new approaches and procedures. In this regard, it is recommended to reduce the number of steps in a given procedure concerning a medical device in order to improve its usability effectiveness among older people [8]. Another important general recommendation for effectively facilitating user tasks for older persons is building redundant cues in the medical device design [16].

Cumulative trauma disorder (CTD) is most prevalent among individuals with occupations requiring the performance of the same task repeatedly. In fact, the affliction rate for CTD disorders appears to be as high as 25% for people performing motion-intensive tasks with their hands [17]. CTD presents a significant degree of threat to healthcare workers, but it can be prevented through actions such as better device design and better work habits reinforced through warning labels and instructions [18].

CTD risk factors are not completely understood, mainly because the onset of symptoms is usually a slow process. Nonetheless, various studies conducted over the years have successfully identified numerous risk factors. For example, risk factors common to handheld medical device applications include poor posture, high and sustained muscle exertion, vibrations, mechanical pressure, sustained posture, and repetitive movements [8]. Some useful guidelines for designing hand-operated devices with respect to CTD are as follows [8, 19]:

- Ensure that gripping surfaces and controls are designed so they enable potential users to keep their hands in a neutral, resting position.
- Provide padding and ergonomically contoured surfaces to reduce concentration of mechanical stresses on the skin and underlying tissues.
- Provide force-assist mechanisms as necessary to decrease the muscle exertion needed to operate a device under consideration.
- Reduce the weight of objects to levels they can easily be picked up or moved.
- Ensure that adequate space is provided for forearm and hand movements to prevent potential users from assuming poor hand postures while carrying out tasks.
- Design hand-operated devices so they will be quite comfortable for individuals with different hand sizes.
- Perform analysis of the range of user hand motion as a basis for determining the dynamic characteristics of controls and handles.
- Design objects in such a way so they can easily be grasped by the entire hand, rather than pinched between fingers and the thumb, in circumstances, when a high degree of precision is not required.
- Shield devices to reduce vibrations they will transmit to users.
- Avoid those designs that will require potential users to exert force continuously.
- Design heavy/awkwardly shaped items so they can easily be grasped or lifted by both hands of users.
- Provide appropriate instructions to all potential users on how to stop the occurrence of CTDs.
- Provide appropriate advisory instructions/visual cues with respect to holding a device.
- Choose appropriate materials for handles that provide a non-slip grip and protection to the hands of users from electrical conduction, vibrations, and cold temperatures.
- Develop operational sequences that prevent the frequent occurrence of a repetitive movement.
- Position work surfaces in such a way that allows forearms to extend at an angle of approximately 90 degrees with respect to the body, with the elbows held at one's side.

## 6. USEFUL DOCUMENTS FOR IMPROVING MEDICAL DEVICE USABILITY

Over the years, many publications on usability and related areas have appeared. Some of these publications that can directly or indirectly be useful for improving medical device usability are listed below.

- Gosbee, J., The Discovery Phase of Medical Device Design: A Blend of Intuition, Creativity, and Science, Medical Device & Diagnostic Industry Magazine, October 1997, pp. 113–118.
- Brown, D., The Challenges of User-Based Design in a Medical Equipment Market, In Field Methods Casebook for Software Design, John Wiley and Sons, New York, 1996, pp. 157–176.
- Gosbee, J., Ritchie, E. M., Human-Computer Interaction and Medical Software Development, Interactions, 1997, Vol. 4, No. 4, pp. 13–18.
- ANSI/AAMI-HE-48, Human Factors Engineering Guidelines and Preferred Practices for the Design of Medical Devices. This standard was developed by the Association for the Advancement of Medical Instrumentation (AAMI) and approved by the American National Standards Institute (ANSI), AAMI, Arlington, Virginia, 1993.
- Hyman, W. A., Errors in the Use of Medical Equipment, In Human Error in Medicine, Lawrence Erlbaum Associates, Inc., Hillsdale, New Jersey, 1994, pp. 327–347.
- Clans, P. L., Gibbons, P. S., Kaihoi, B. H., Usability Laboratory: A New Tool for Process Analysis at the Mayo Clinic, Proceedings of the Healthcare Information Management Systems Society Conference, 1997, pp. 149–159.
- Volaitis, L. E., Chou, R. S., Wiklund, M., Consumers' Expectations of the Weights of Portable Communications Devices, Proceedings of the 42nd Human Factors and Ergonomics Society Annual Meeting, 1997, pp. 140–146.
- Navai, M., Guo, X., Caird, J. K., Dewar, R. E., Understanding of Prescription Medical Labels as a Function of Age, Culture, and Language, Proceedings of the Human Factors and Ergonomics Society Conference, 2000, pp. 1487–1491.

- Morrow, D. G., Leirer, V. O., Andrassy, J. M., Using Icons to Convey Medication Schedule Information, Applied Ergonomics, Vol. 27, 1996, pp. 267–275.
- Seagull, F. J., Sanderson, P. M., Anesthesia Alarms in Context: An Observational Study, Human Factors, Vol. 43, 2001, pp. 66–78.
- Burgess-Limerick, R., Mon-Williams, M., Coppard, V. L., Visual Display Height, Human Factor, Vol. 42, 2000, pp. 140–150.
- Chaffin, D. B., Faraway, J. J., Zhang, X., Woolley, C., Stature, Age, and Gender Effects on Reach Motion Postures, Human Factors, Vol. 42, 2000, pp. 408–420.
- Designer's Handbook: Medical Electronics, Canon Communications, Inc., Santa Monica, California, 1995.
- Hix, D., Hartson, R., Developing User Interfaces: Ensuring Usability Through Product and Process, John Wiley and Sons, New York, 1993.
- Wiklund, M., Editor, Usability in Practice: How Companies Develop User-Friendly Products, Academic Press, Inc., Cambridge, Massachusetts, 1994.


## 7. PROBLEMS

1. List at least seven important characteristics of the user population that should be taken into consideration during the medical device design.
2. List at least 10 medical devices with high incidence of user error.
3. Describe a general approach for developing effective user interfaces of medical devices.
4. Discuss the following factors with respect to medical device use environments:
   - Mental workload
   - Physical workload
   - Motion and vibration
5. List areas on which information is necessary to understand a medical device's use accurately and completely.
6. List at least 10 guidelines for making medical device interfaces more user-friendly.
7. Discuss at least five of the guidelines from Question 6 in detail.
8. Discuss cumulative trauma disorder (CTD) implications in medical devices.
9. What are the important factors that must be considered in designing medical devices for use by older people?
10. List at least 10 useful guidelines for designing hand-operated devices with respect to CTD.

## 8. REFERENCES


1. Fortune Magazine (USA), May 17, 1993, pp. 10.
2. Information Infrastructure for Healthcare, Advanced Technology Program, National Institute of Standards and Technology (NIST), U.S. Department of Commerce, Washington, D.C., 2003, Available online at http://www.atp.nist.gov/www/press/iih9703.htm.
3. Garmer, K., Liljegren, E., Osvalder, A. L., Dahlman, S., Arguing for the Need of Triangulation and Iteration When Designing Medical Equipment, Journal of Clinical Monitoring and Computing, Vol. 17, 2002, pp. 105–114.
4. Sawyer, D., Do It By Design: An Introduction to Human Factors in Medical Devices, Center for Device and Radiological Health, Food and Drug Administration, Washington, D.C., 1997.
5. Hayman, W. A., Errors in the Use of Medical Equipment, in Human Error in Medicine, edited by M. S. Bogner, Lawrence Erlbaum Associates, Inc., New York, 1994, pp. 327–347.
6. Obradovich, J. H., Woods, D. D., Users as Designers: How People Cope with Poor HCI Design in Computer-Based Medical Devices, Human Factors, Vol. 38, 1996, pp. 40–46.
7. Medical Device Use-Safety: Incorporating Human Factors Engineering into Risk Management, Draft Guidance Document, Center for Devices and Radiological Health, Food and Drug Administration, Washington, D.C., 2000.
8. Wiklund, M. E., Medical Device and Equipment Design: Usability Engineering and Ergonomics, Interpharm Press, Inc., Buffalo Grove, Illinois, 1995.
9. Medical Device Reporting (MDR) System, Center for Devices and Radiological Health (CDRH), Food and Drug Administration, Washington, D.C., 1991.
10. Salvemini, A. J., Challenge for User-Interface Designers of Telemedicine Systems, Telemedicine Journal, Vol. 5, No. 2, 1999, pp. 10–15.

11. Wiklund, M. E., Making Medical Device Interfaces More User Friendly, Medical Device & Diagnostic Industry (MDDI) Magazine, May 1998, pp. 177–184.
12. Czaja, S, Special Issue Preface, Human Factors, Vol. 32, No. 5, 1990, pp. 505.
13. CPSC Publication No. 702, Product Safety and the Older Consumer: What Manufacturers/Designers Need to Consider, Consumer Product Safety Commission (CPSC), Washington, D.C., 1988.
14. Small, A., Design for Older People, in Handbook of Human Factors, edited by G. Salvendy, John Wiley and Sons, New York, 1987, pp. 125–140.
15. Czaja, S., Clark, M., Weber, R., Computer Communication Among Older Adults, Proceedings of the Human Factors Society 34th Annual Meeting, 1990, pp. 304–309.
16. Koncelik, J., Aging and the Product Environment, Scientific and Academic Additions, Inc., Florence, Kentucky, 1982.
17. Armstrong, T., Radwin, R. G., Hansen, D. J., Repetitive Trauma Disorders: Job Evaluation and Design, Human Factors, Vol. 28, No. 3, 1986, pp. 325–330.
18. Hebert, L., Living with CTD, IMPACC, Inc., Bangor, Maine, 1990.
19. Putz-Anderson, V., Cumulative Trauma Disorders, Taylor & Francis, Inc., New York, 1988.

# APPENDIX

# BIBLIOGRAPHY: LITERATURE ON USABILITY ENGINEERING

## A.1 INTRODUCTION

Over the years, many publications on usability engineering and related areas have appeared in the form of journal papers, conference proceeding papers, books, technical reports, and so on. This appendix presents a comprehensive list of the publications.

The period covered by the listing is from 1975-2002. The main objective of this listing is to provide readers with sources of additional information on usability engineering and associated areas.

## A.2 PUBLICATIONS


1. Abernethy, C. N., Human-Computer Interface Standards: Origins, Organizations and Comment, International Review of Ergonomics, Vol. 2, 1988, pp. 31–54.
2. Adler, P. S., Winograd, T. A., Editors, Usability: Turning Technologies into Tools, Oxford University Press, New York, 1992.
3. Adolf, J. A., Holden, K. L., Touchscreen Usability in Microgravity, Proceedings of the Conference on Human Factors in Computing Systems, 1996, pp. 67–68.
4. Ames, A. L., Users First: An Introduction to Usability and User-Centered Design and Development for Technical Information and Products, Proceedings of the IEEE International Professional Communication Conference, 2001, pp. 135–140.
5. Anderson, J., et. al., Integrating Usability Techniques into Software Development, IEEE Software, Vol. 18, No. 1, 2001, pp. 46–53.
6. Anderson, J. R., Jeffries, R., Novice LISP Errors: Undetected Losses From Working Memory, Human Computer Interaction, Vol. 1, 1985, pp. 107–131.
7. Andre, T. S., et al, User Action Framework: A Reliable Foundation for Usability Engineering Support Tools, International Journal of Human-Computer Studies, Vol. 54, No. 1, 2001, pp. 107–136.
8. Apple Computer, Human Interface Guidelines: The Apple Desktop Interface, Addison Wesley Publishing, Reading, Massachusetts, 1987.
9. Asahi, T., et al., Usability Testing Method Employing the "Trouble Model", Proceedings of the Human Factors Society Conference, 1990, pp. 1233–1237.
10. Ashlund, S. L., Horwitz, K. J., Usability Improvements in Lotus CC: Mail for Windows, Proceedings of the Conference on Human Factors in Computing Systems, 1996, pp. 481–488.
11. Atyeo, M., User Interface Architecture, BT Technology Journal, Vol. 6, No. 4, 1988, pp. 17–23.
12. Aykin, N., Software Reuse: A Case Study on Cost-Benefit of Adopting a Common Software Development Tool, in Cost Justifying Usability Edited by R. G. Bias and D. J. Mayhew, Morgan Kaufmann Publishers, San Francisco, 1994, pp. 177–202.
13. Aykin, N. A., et al, Designing Operations Systems Interfaces that are Easy to Use, AT & T Technical Journal, Vol. 73, 1994, pp. 14–21.
14. Badre, A., Jacobs, A., Usability, Aesthetics, and Efficiency: An Evaluation In a Multimedia Environment, Proceedings of the International Conference on Multimedia Computing & Systems,


ISBN: 1-58883-084-5 / $35.00


*Engineering Usability: Fundamentals, Applications, Human Factors, and Human Error*
Authored by B. S. Dhillon

1999, pp. 103–106.

15. Baerentsen, K. B., Slavensky, H., A Contribution to the Design Process, Communications of the ACM, Vol. 42, No. 5, 1999, pp.73–77.
16. Bailey, J. E., Pearson, S. W., Development of a Tool For Measuring and Analyzing Computer User Satisfaction, Management Science, Vol. 29, No. 6, 1983, pp.519–529.
17. Bailey, R. W., Human Performance Engineering: Using Human Factors/Ergonomics to Achieve Computer System Usability, Prentice Hall, Englewood Cliffs, New Jersey, 1989.
18. Bailey, W. A., Knox, S. T., Lynch, E. F., Effects of Interface Design Upon User Productivity, Proceedings of the Conference on Human Factors in Computing Systems, 1988, pp. 207–212.
19. Barker, R. T., Biers, D. W., Software Usability Testing: Do Use Self-Consciousness and the Laboratory Environment Make Any Difference?, Proceedings of the Human Factors and Ergonomics Society Conference, 1994, pp. 1131–1134.
20. Beckner, S. A., Mottay, F. E., Global Perspective on Web Site Usability, IEEE Software, Vol. 18, No. 1, 2001, pp. 54–61.
21. Benbunan-Fich, R., Using Protocol Analysis to Evaluate the Usability of a Commercial Web Site, Information Management, Vol. 39, No. 2, 2001, pp. 151–163.
22. Benimoff, N. I., Whitten, W. B., Human Factors Approaches to Prototyping and Evaluating User Interfaces, AT and T Technical Journal, Vol. 68, No. 5, 1989, pp. 44–55.
23. Bennett, J., et al., editors, Visual Display Terminals: Usability Issues and Health Concerns, Prentice Hall, Englewood Cliffs, New Jersey, 1984.
24. Bennett, J., Managing to Meet Usability Requirements, In Visual Display Terminals: Usability Issues and Health Concerns edited by J. Bennett, D. Case, J. Sandelin, and M. Smith, Prentice-Hall, Englewood Cliffs, New Jersey, 1984, pp. 161–184.
25. Best, P. S. et al., Evaluating Thermal and Image Intensification Night Vision Devices for the Ground Environment: Human Factors and Usability Issues, Proceedings of the IEEE National Aerospace and Electronics Conference, 1998, pp. 295–301.
26. Bias, R. G., Mayhew, D. J., editors, Cost-Justifying Usability, Academic Press, Boston, 1994.
27. Bias, R. G., Wherefore Cost Justification of Usability: Pay Me Now or Pay Me Later: But How Much?, in Cost Justifying Usability edited by R. G. Bias and D. J. Mayhew, Morgan Kaufmann Publishers, San Francisco, 1994, pp. 3–8.
28. Blatt, L. A., Knutson, J. F., User Interface Design Guidance Systems, Proceedings of the Human Factors and Ergonomics Society Conference, 1993, pp. 1038–1042.
29. Bloomer, S. A., Wolfe, S. J., Hiser Group: Pioneering Usability and User Interface Design in Australia, Proceedings of the Conference on Human Factors in Computers Systems, 1996, pp. 141–142.
30. Bodker, S., Through the Interface: A Human Activity Approach to User Interface Design, Lawrence Erlbaum Associates Publishers, Hillsdale, New Jersey, 1991.
31. Borgholm, T., Madsen, K. H., Cooperative Usability Practices, Communications of the ACM, Vol. 42, No. 5, 1999, pp.91–97.
32. Bowers, A., Snyder, H. L., Concurrent Versus Retrospective Verbal Protocol For Comparing Window Usability, Proceedings of the Human Factors Society Conference, 1990, pp. 1270–1274.
33. Brooks, F., No Silver Bullet: Essence and Accidents of Software Engineering, IEEE Computer, Vol. 20, No. 4, 1987, pp. 10–19.
34. Brooks, R., Justifying Prepaid Human Factors for User Interfaces, in Cost Justifying Usability edited by R. G. Bias and D. J. Mayhew, Morgan Kaufmann Publishers, San Francisco, 1994, pp. 273–286.
35. Butler, K. A., Connecting Theory and Practice: A Case Study of Achieving Usability Goals, Proceedings of the Conference on Human Factors in Computing Systems, 1985, pp. 85–88.
36. Butler, K. A., Jacob, R. J. K., John, B. E., Introduction & Overview to Human-Computer Interaction, Proceedings of the Conference on Human Factors in Computing Systems, 1995, pp. 345–346.
37. Butler, K. A., Usability Engineering Turns Ten, Interactions, Vol. 3, No. 1, 1996, pp. 59–75.
38. Buur, I., Bagger, K., Replacing Usability Testing with user Dialogue, Communications of the ACM, Vol. 42, No. 5, 1999, pp. 63–66.
39. Buur, J., Bodker, S., From Usability Lab. To "Design Collaboratorium": Reframing Usability Practice, Proceedings of the Conference on Designing Interactive Systems: Processes, Practices,

Methods, and Techniques, 2001, pp. 297–307.
40. Buxton, B., Smoke and Mirrors, Byte, Vol. 15, No. 7, 1990, pp. 205–210.
41. Canavesio, F., Marion, R., Optimization of a Human Interface to Access Image Data Bases, CSELT Technical Reports, Vol. 17, No. 4, 1989, pp. 259–263.
42. Card, S. K., Moran, T. P., Newell, A., The Psychology of Human-Computer Interaction, Lawrence Erlbaum Associates Publishers, Hillsdale, New Jersey, 1983.
43. Carey, J. W., Usability Guidelines for Simulation Package Interfaces, Simulation Series, Vol. 21, No. 4, pp. 147–151.
44. Carroll, J. M., Carrithers, C., Blocking Learner Errors in a Training Wheels Systems, Human Factors, Vol. 26, No. 4, 1984, pp. 377–389.
45. Carroll, J. M., Kay, D. S., Prompting Feedback and Error Correction in the Design of a Scenario Machine, Proceedings of the Conference on Human Factors in Computing Systems, 1985, pp. 149–154.
46. Carroll, J. M., Making Use: Scenario-Based Design of Human-Computer Interactions, MIT Press, Cambridge, Massachusetts, 2000.
47. Carroll, J. M., Rosson, M. B., Usability Specifications as a Tool in Interative Development, in Advances in Human Computer Interaction, edited by R. Hartson, Ablex Publishing Corporation, Norwood, New Jersey, 1985, pp. 1–28.
48. Catani, M. B., Biers, D. W., Usability Evaluation and Prototype Fidelity: Users and Usability Professionals, Proceedings of the Human Factors Society Conference, 1998, pp. 1331–1335.
49. Caulton, D. A., Relaxing the Homogenity Assumption In Usability Testing, Behavior & Information Technology, Vol. 20, No. 1, 2001, pp. 1–7.
50. Chapanis, A., Evaluating Usability, in Human Factors for Informatics Usability, edited by B. Shackel and S. Richardson, Cambridge University Press, London, 1991, pp. 359–398.
51. Chen, C., Wilcox, L., Introduction to the Human Factors and Usability Issues Minitrack, Proceedings of the Hawaii, International Conference on System Sciences, 1999, pp. 68–72.
52. Cherry, J. M., Jackson, S. R., Online Help: Effects of Content and Writing Style on User Performance and Attitudes, IEEE Transactions on Professional Communications, Vol. 32, No. 4, 1989, pp. 294–299.
53. Chignell, M. H., Valdez, J. F., Truncated Experiments and Fields Experiments in Usability Engineering, Proceedings of the IEEE International Conference on Systems, Man, and Cybernetics, 1991, pp. 1349–1353.
54. Christ, R. E., Review and Analysis of Color Coding Research for Visual Displays, Human Factors, Vol. 17, 1975, pp. 542–570.
55. Chrush, M., Seven Great Myths of Usability, Interactions, Vol. 3, No. 1, 2000, pp. 13–16.
56. Coggman, L., Cohen, S. J., Usability Engineering from the Supplier's Point of View, IEE Colloquium (Digest), No. 73 , 1995, pp. 6/1–6/4.
57. Conklin, P. F., Bringing Usability Effectively Into Product Development, in Human-Computer Interface Design: Success Stories, Emerging Methods, Real-World Context, edited by M. Rudisill, C. Lewis, P. Polson, and T. McKay, Morgan Kaufmann Publishers, San Francisco, 1991, pp. 367–385.
58. Conn, A. P., Time Affordances the Time Factor in Diagnostic Usability Heuristics, Proceedings of the Conference on Human Factors in Computing Systems, 1995, pp. 186–193.
59. Constantine, L. L., Lockwood, L. A. D., Software for Use: A Practical Guide to the Models and Methods of Usage-Centered Design, Addison-Wesley Publishing, Reading, Massachusetts, 1999.
60. Cope, M. E., Uliano, K. C., Cost-Justifying Usability Engineering: A Real World Example, Proceedings of the Human Factors and Ergonomics Society Conference, 1995, pp. 263–267.
61. Costabile, M. F., Usable Multimedia Applications, Proceedings of the International Conference on Multimedia Computing & Systems, 1999, pp. 124–127.
62. Cox, M. E., O'Neal, P., Pendley, W. L., UPAR Analysis: Dollar Measurement of a Usability Indicator for Software Products, in Cost Justifying Usability edited by R. G. Bias and D. J. Mayhew, Morgan Kaufmann Publishers, San Francisco, 1994, pp. 143–158.
63. Crossby, O., You are a What? Usability Engineer, Occupational Outlook Quarterly, Vol. 44, No. 4, 2001, pp. 10–13.
64. Curtis, B., Nielsen, J., Applying Discount Usability Engineering, IEEE Software, Vol. 12, 1995, pp.

98–100.

65. Desurvire, H., Thomas, J. C., Enhancing the Performance of Interface Evaluators Using Non-Empirical Usability Methods, Proceedings of the Human Factors and Ergonomics Society Conference, 1993, pp. 1132–1136.
66. DeYoung, L., Interacting with Users in the Design Process, Proceedings of the IEEE Computer Society Conference, 1991, pp. 500–504.
67. Dick, D., Strategies for Usability Putting ISO Standards to Practice, Proceedings of the Society for Technical Communication Annual Conference, 2001, pp. 391–394.
68. Dieli, M., Usability Process: Working with Iterative Design Principles, IEEE Transactions on Professional Communications, Vol. 32, No. 4, 1989, pp. 272–278.
69. Dillon, A., Group Dynamics Meet Cognition: Combining Sociotechnical Concepts and Usability Engineering in the Design of Information Systems, in the New Socio Tech: Graffitti on the Long Wall,, edited by E. Coakes, Springer-Verlag, New York, 2000, pp. 119–126.
70. Dillon, A., Morris, M., User Acceptance of New Information Technology: Theories and Models, Annual Review of Information Science and Technology, Vol. 31, 1996, pp. 3–32.
71. Dillon, A., Sweeney, S., Maguire, M., A Survey of Usability Engineering Within the European IT Industry: Current Practice and Needs, Proceedings of the Conference on Human-Computer Interactions (HCI'93), 1993, pp. 81–94.
72. Dolan, W. R., Dumas, J. S., A Flexible Approach to Third-Party Usability, Communications of the ACM, Vol. 42, No. 5, 1999, pp.83–85.
73. Downey, L. L., Laskowski, S. J., Buie, E. A., Hartson, H. R., Usability Engineering: Industry-Government Collaboration for System Effectiveness and Efficiency, SIGCHI, Vol. 28, No. 4, 1996, pp. 66–67.
74. Downey, L. L., Laskowski, S. J., Buie, E. A., Hefley, W. E., Usability Engineering 2: Measurement and Methods, SIGCHI, Vol. 30, No. 2, 1998, pp. 10–14.
75. Dray, S. M. Karat, C. M., Human Factors Cost Justification of an Internal Development Project, in Cost Justifying Usability, edited by R. G. Bias and D. J. Mayhew, Morgan Kaufmann Publishers, San Francisco, 1994, pp. 111–122.
76. Dumas, J. S., Redish, J. C., A Practical Guide to Usability Testing, Ablex Publishing Corporation, Norwood, New Jersey, 1993.
77. Eason, K. D., Harker, S., The Supplier's Role in the Design of Products for Organizations, The Computer Journal, Vol. 31, No. 5, 1999, pp.344–348.
78. Ebling, M., John, B. E., On the Contributions of Different Empirical Data in Usability Testing, Proceedings of the Conference on Designing Interactive Systems: Processes, Practices, Methods, and Techniques, 2001, pp. 289–296.
79. Edwards, A., Extra-Ordinary Human-Computer Interactive: Interfaces for Users with Disabilities, Cambridge University Press, Cambridge, U.K., 1995.
80. Ehrlich, K., Roh, J. A., Cost Justification of Usability Engineering: A Vendor's Perspective, in Cost Justifying Usability edited by R. G. Bias and D. J. Mayhew, Morgan Kaufmann Publishers, San Francisco, 1994, pp. 73–109.
81. Farber, C. G., Software Instllation In a Distributed Computing Environment: Human Factors Issues and Methods, Proceedings of the Human Factors and Ergonomics Society Conference, 1993, pp. 277–281,
82. Federico, M., Usability Evaluation of a Spoken Data-Entry Interface, Proceedings of the International Conference on Multimedia Computing, 1999, pp. 726–731.
83. Feldkamp, F., Heinrich, M., Meyer-Gramann, K. D., SyDeR-System Design for Reusability, Artificial Intelligence for Engineering Design, Analysis & Manufacturing: Aiedarn, Vol. 12, No. 4, 1998, pp. 373–382.
84. Ferre, Z., Juristo, N., Windl, H., Constantine, L., Usability Basics for Software Developers, IEEE Software, Vol. 18, No. 1, 2001, pp. 22–29.
85. Flamm, L. E., Usability Testing: Where Does It Belong in Software Design?, Proceedings of the IEEE International Conference on Systems, Man, and Cybernetics, 1989, pp. 235–236.
86. Flanagan, G. A., Ranch, T. L., Usability Management Maturity, Part I: Self Assessment-How Do You Stack Up?, Proceedings of the Conference on Human Factors in Computing Systems, 1995, pp.

336–339.
87. Flohrer, W., Human Factors: Design of Usability, Proceedings of the IEEE Conference on VLSI and Computer Peripherals, 1989, pp. 108–113.
88. Frederick, M., Usability Evaluation of a Spoken Data-Entry Interface, Proceedings of the International Conference on Multimedia Computing & Systems, 1999, pp. 726–731.
89. Frokjaer, E., Hertzum, M., Hornbaek, K., Measuring Usability: Are Effectiveness, Efficiency, and Satisfaction Really Correlated?, Proceedings of the Conference on Human Factors in Computing Systems, 2000, pp. 345–352.
90. Fu. L., Salvendy, G., Turley, L., Who Finds What in Usability Evaluation, Proceedings of the Human Factors Society Conference, 1998, pp. 1341–1345.
91. Gabbard, J. L., et al., Usability Engineering: Domain Analysis Activities for Augmented Reality Systems, Proceedings of the International Society for Optical Engineering, 2002, pp. 445–457.
92. Gaines, B. R., Chen, L. L., Shaw, M. L. G., Modeling the Human Factors of Scholarly Communities Supported Through the Internet and World Wide Web. Journal of the American Society for Information Science, Vol. 48, No. 11, 1997, pp. 987–1003.
93. Galitz, W. O., The Essential Guide to User Interface Design: An Introduction to GUI Design Principles and Techniques, John Wiley and Sons, New York, 1996.
94. Gallters, J., Sutcliffe, A., Minocha, S., An Impact Analysis Method for Safety Critical User Interface Design, ACM Transactions on Computer-Human Interaction, Vol. 6, 1999, pp. 341–369.
95. Gardner, J., Strengthening the Focus on User's Working Practices, Vol. 42, No. 5, 1999, pp.79–82.
96. George, H., The Good Usability Handbook, McGraw-Hill Book Company, New York, 1996.
97. Gillan, D. J., Bias, R. G., The Interface Between Human Factors and Design, Proceedings of the Human Factors Society Annual Conference, 1992, pp. 443–447.
98. Gillan, D. J., Breedin, S. D., Cooke, N. J., Network and Multidimensional Representations of the Declarative Knowledge of Human-Computer Interface Design Experts, International Journal of Man-Machine Studies, Vol. 36, 1992, pp. 587–615.
99. Gillan, D. J., Cooke, N. J., Making Usability Data More Usable, Proceedings of the Human Factors Society Conference, 1998, pp. 300–304.
100. Gnisci, A., Papa, F., Spedaletti, S., Usability Aspects, Socio-Relational Context and Learning Performance in the Virtual Classroom: A Laboratory Experiment, Behaviour and Information Technology, Vol. 18, No. 6, 1999, pp. 431–443.
101. Gould, J. D., Boies, S. J., Human Factors Challenges in Creating Principle Support Office: The Speech Filing System Approach, ACM Trans. Of Office Information Systems, Vol. 1, No. 4, 1983, pp. 273–298.
102. Gould, J. D., Boies, S. J., Ukelson, J., How to Design Usable Systems, in the Handbook of Human - Computer Interaction, edited by M. G., Helander, K., Thomas, Prabhu, P. V., Elsevier Science Publishers, Amsterdam, 1997, pp. 236–245
103. Gould, J. D., Lewis, C. H., Designing for Usability: Key Principles and What Designers Think, Communications of the ACM, Vol. 28, No. 3, 1985, pp. 300–311.
104. Govindaraja, M., Mital, A., Enhancing the Usability of Consumer Products Through Manufacturing: Part I—Developing the Usability-Manufacturing Attribute Linkages for A Hybrid Bicycle, International Journal of Industrial Engineering, Vol. 7, No. 1, 2000, pp. 33–43.
105. Govindaraju, M., Mital, A., Enhancing the Usability of Consumer Products Through Manufacturing: Part II—Development of Usability and Design Checklists for a Hybrid Bicycle, International Journal of Industrial Engineering, Vol. 7, No. 1, 2000, pp. 44–51.
106. Granic, A., Glavinic, U., Usability Evaluation Issues for Computerized Educational Systems, Proceedings of the Mediterranean Electrotechnical Conference, 2002, pp. 558–562.
107. Gray, W. D., Salzman, M. C., Damanged Merchandise? A Review of Experiments That Compare Usability Evaluation Methods, Human-Computer Interaction, Vol. 13, 1998, pp. 203–261.
108. Greehalgh, C. M., Evaluating the Network and Usability Characteristics of Virtual Reality Conferencing, BT Technology Journal, Vol. 15, No. 4, 1997, pp. 101–119.
109. Green, W. S., Jordan, P. W., Pleasure with Products: Beyond Usability, Taylor and Francis, New York, 2001.
110. Grudin, J., Interface: An Evolving Concept, Communications of the ACM, Vol. 36, 1993, pp.

110–119.

111. Guindon, R., Designing the Design Process: Exploiting Opportunistic Thoughts, Human-Computer Interaction, Vol. 5, 1990, pp. 305–344.
112. Gunn, C., Example of Formal Usability Inspections in Practice at Hewlett-Packard Company, Proceedings of the Conference on Human Factors in Computing Systems, 1995, pp. 103–104.
113. Hackos, J. T., Redish, J. C., User and Task Analysis for Interface Design, John Wiley and Sons, New York, 1998.
114. Hall, S. K., Cockerham, K. J., Rhodes, D. J., Applying Human Factors in Graphical Operator Interfaces, Proceedings of the IEEE Annual Pulp and Paper Industry Technical Conference, 2001, pp. 241–246.
115. Hamlet, D., Mason, D., Woit, D., Theory of Software Reliability Based on Components, Proceedings of the International Conference on Software Engineering, 2001, pp. 361–370.
116. Harrison, M. C., Henneman, R. L., Blatt, L. A., Design of a Human Factors Cost-Justification Tool, in Cost Justifying Usability edited by R. G. Bias and D. J. Mayhew, Morgan Kaufmann Publishers, San Francisco, 1994, pp. 203–241.
117. Haubner, P. J., Ergonomics in Industrial Product Design, Ergonomics, Vol. 33, No. 4, 1990, pp. 477–485.
118. Heasly, C. C., et al, High Definition Systems Usability Test Tool (HUTT): A Human-Computer Interface Prototyping Tool, Proceedings of the Human Factors and Ergonomics Society Conference, 1993, pp. 1041–1043.
119. Henninger, S., Lu, C., Candace, F., Using Organizational Learning Techniques to Develop Context-Specific Usability Guidelines, Proceedings of the Conference on Designing Interactive Systems, 1997, pp. 129–136.
120. Hewitt, K. L., et al., Applying Competitive Usability Advantage, Proceedings of the Human Factors Society Conference, 1998, pp. 1611–1615.
121. Hinderer, D., Challenges in Participant Recruiting for Usability Testing, Proceedings of the IEEE International Professional Communication Conference, 2000, pp. 417-426.
122. Hix, D., Hartson, H. R., Developing User Interfaces: Ensuring Usability Through Product and Process, John Wiley and Sons, New York, 1993.
123. Hoffman, J., Cook, C. A., Designing for Usability with COTS? How Useful is a Style Guide?, Proceedings of the Human Factors Society Conference, 1998, pp. 1295–1299.
124. Holcomb, R., Tharp, A. L., What Users Say About Software Usability, International Journal of Human-Computer Interaction, Vol. 3, No. 1, 1991, pp. 49–78.
125. Howard, S., Hammond, J., Lindgaard, G., editors, Human-Computer Interaction: Interact'97, Chapman and Hall, London, 1997.
126. Hyman, W. A., Errors in the Use of Medical Equipment, In Human Error in Medicine, edited by M. S. Bogner, Lawrence Erlbaum Associates, Inc., Hillsdales, New Jersey, 1994, pp. 327–347.
127. ISO DIS 13407 (1997), User Centered Design Process for Interactive Systems, International Standardization Organization (ISO), Geneva, Switzerland.
128. ISO DIS 9241-11 (1996), Ergonomics Requirements For Office Work With Visual Display Terminals (VDTs): Part II—Guidance on Usability, International Standardization Organization (ISO), Geneva, Switzerland.
129. Ives, B., Olson, M. H., Baroudi, J. J., The Measurement of User Information Satisfaction, Communication of the ACM, Vol. 26, No. 10, 1983, pp.785–793.
130. Jacob, R. J. K., Using Formal Specifications in the Design of a Human-Computer Interface, Communications of the ACM, Vol. 26, No. 4, 1983, pp. 259–264.
131. Jacobsen, N. E., Morten, H., John, B. E., Evaluator Effect in Usability Studies: Problem Detection and Severity Judgements, Proceedings of the Human Factors Society Conference, 1998, pp. 1336–1340.
132. Jambon, F., Girard, P., Ait-Ameur, Y., Interactive System Safety and Usability Enforced with the Development Process, Proceedings of the Engineering for Human-Computer Interaction Conference, 2001, pp. 61–76.
133. Jordan, P. W., An Introduction to Usability, Taylor and Francis, London, 1998.
134. Jordan, P. W., Thomas, B., Weerdmeester, B., Editors, Usability Evaluation in Industry, Taylor and Francis, London, 1996.

135. Jurgen, K. B., Carroll, J. M., Rosson, M. B., Singley, M. K., Comparative Usability Evaluation: Critical Incidents and Critical Threads, Proceedings of the Conference on Human Factors in Computing Systems, 1994, pp. 245–251.
136. Kalawsky, R. S., VRUSE: A Computerized Diagnostic Tool: For Usability Evaluation of Virtual/ Synthetic Environment, Applied Ergonomics, Vol. 30, No. 1, 1999, pp. 11–25.
137. Karat, C., Cost-Benefit Analysis of Usability Engineering Techniques, Proceedings of the Human Factors Society Conference, 1990, pp. 839–843.
138. Karat, C., Cost-Justifying Usability, in the Handbook of Human-Computer Interaction edited by M. G. Helander, T. K. Landauer, and P. V., Prabhu, North-Holland Publishing Company, Amsterdam, 1997, pp. 767–781.
139. Karat, C., Usability Engineering in Dollars and Cents, IEEE Software, Vol. 10, No. 3, 1993, pp. 88–89.
140. Karat, C. M., A Business Case Appraoch to Usability Cost Justification, in Cost Justifying Usability edited by R. G. Bias and D. J. Mayhew, Morgan Kaufmann Publishers, San Francisco, 1994, pp. 45–70.
141. Karat, J., Dayton, T., Practical Education for Improving Software Usability, Proceedings of the Conference on Human Factors in Computing Systems, 1995, pp. 162–169.
142. Kellogg, W., A., Breen, T. J., Evaluating User and System Models: Applying Scaling Techniques to Problems In Human-Computer Interaction, Proceedings of the Conference on Graphics Interface, 1987, pp. 303–309.
143. Kelly, J. F., et al., Extending User-Centered Methods Beyond Interface Design to Functional Definition, Proceedings of the Human Factors and Ergonomics Society Conference, 1996, pp. 343–347.
144. Kerton, B., Introducing Usability at London Life Insurance Company: A Process Perspective, Proceedings of the Conference on Human Factors in Computing Systems, 1997, pp. 300–304.
145. Kirakowski, J., Cierlik, B., Measuring the Usability of Web Sites, Proceedings of the Human Factors Society Conference, 1998, pp. 424–428.
146. Kohoutek, H. J., Reliability Engineering Concepts in Design for Usability, Quality and Reliability Engineering International, Vol. 10, No. 2, 1994, pp. 133–141.
147. Kwahk, J., et al, Selection and Classification of the Usability Attributes for Evaluating Consumer Electronic Products, Proceedings of the Human Factors and Ergonomics Society Conference, 1997, pp. 432–436.
148. Landauer, T. K., The Trouble with Computers: Usefulness, Usability Productivity, MIT Press, Cambridge, Massachusetts, 1995.
149. Lauesen, S., Usability Engineering In Industrial Practice, in Human-Computer Interaction, edited by S. Howard, J. Hammond and G. Lindgaard, Chapman and Hall, London, 1997, pp. 15–22.
150. Lecerof, A., Paterno, F., Automatic Support for Usability Evaluation, IEEE Transactions on Software Engineering, Vol. 24, No. 10, 1998, pp. 863-–888.
151. Lederer, A. L., Prassad, J., Nine Management Guidelines for Better Cost Estimating, Communications of the ACM, Vol. 35, No. 2, 1992, pp. 51–59.
152. Lelouche, R., Page, M., Save on Time and Effort: Application of Discount Usability Engineering to the Design of Internet-Based Learning Environments, International Journal of Continuing Engineering Education, Vol. 11, No. 1–2, 2001, pp. 35–46.
153. Lewis, C., Norman, D. A., Designing for Error, in User-Centered System Design edited by D. A., Norman and S. Draper, Lawrence Erlbaum Associated Publishers, Hillsdale, New Jersey, 1986, pp. 362–376.
154. Lewis, J.R., IBM Computer Usability Satisfaction Questionnaires: Psychometric Evaluation and Instructions for Use, International Journal of Human-Computer Interaction, Vol. 7, No. 1, 1995, pp. 57-78.
155. Lewis, J. R., Psychometric Evaluation of the Post-Study System Usability Questionnaire: the PSSUQ, Proceedings of the Human Factors Society Conference, 1993, pp. 1259–1263.
156. Lim, K. Y., Structured Task Analysis: An Instantiation of the MUSE Method for Usability Engineering, Interacting with Computers, Vol. 8, No. 1, 1996, pp. 31–50.
157. Lim, K. Y., The MUSE: Method for Usability Engineering, Cambridge University Press, New York,

1994.
158. Lin, H. X., Choong, Y. Y., Salvendy, G., Proposed Index of Usability: A Method For Comparing the Relative Usability of Different Software Systems, Behavior and Information Technology, Vol. 16, No. 4–5, 1997, pp. 267–278.
159. Lindgaard, G., Usability Testing and System Evaluation: A Guide for Designing Useful Computer Systems, Chapman and Hall, London, 1994.
160. Lund, A. W., Tschirgi, J. E., Designing for People: Integrating Human Factors Into the Product Realization Process, IEEE Journal on Selected Areas in Communications, Vol. 9, No. 4, 1991, pp. 496–500.
161. Lyons, K., Starner, T., Mobile Capture for Wearable Computer Usability Testing, Proceedings of the International Symposium on Wearable Computers, 2001, pp. 69–76.
162. MacIntyre, F., Estep, K. W., Sieburth, J. M., Cost of User-Friendly Programming, Journal of Forth Application and Research, Vol. 6, No. 2, 1990, pp. 103–115.
163. Macmichael, R. A., Seven Factors to Consider When Redesigning Your Site, IT Professional, Vol. 3, No. 4, 2001, pp. 35–37.
164. Madsen, K. H., The Diversity of Usability Practices, Communications of the ACM, Vol. 42, No. 5, 1999, pp. 45–50.
165. Mander, R., Smith, B., Web Usability for Dummies, Hungry Minds, New York, 2002.
166. Mania, K., et al., Usability Evaluation Techniques for Virtual Reality Technologies, Proceedings of the Annual International Symposium on Virtual Reality, 2002, pp. 299–303.
167. Mantei, M. M., Theorey, T. J., Cost/Benefit Analysis For Incorporating Human Factors in the Software Life Cycle, Communications of the ACM, Vol. 31, No. 4, 1988, pp.428–439.
168. Mauro, C. L., Cost-Justifying Usability in a Contractor Company, in Cost Justifying Usability edited by R. G. Bias and D. J. Mayhew, Morgan Kaufmann Publishers, San Francisco, 1994, pp. 123–142.
169. Mayhew, D. J, Bias, R. G., Organizational Inhibitors and Facilitators, in Cost Justifying Usability edited by R. G. Bias and D. J. Mayhew, Morgan Kaufmann Publishers, San Francisco, 1994, pp. 297–318.
170. Mayhew, D. J., Cost-Benefit Analysis of Upgrading Computer Hardware, in Cost Justifying Usability edited by R. G. Bias and D. J. Mayhew, Morgan Kaufmann Publishers, San Francisco, 1994, pp. 159–175.
171. Mayhew, D. J., Mantei, M., A Basic Framework for Cost-Justifying Usability Engineering, in Cost Justifying Usability edited by R. G. Bias and D. J., Mayhew, Morgan Kaufmann Publishers, San Francisco, 1994, pp. 9–43.
172. McClelland, I. L., Bringham, F. R., Marketing Ergonomics: How Should Ergonomics be Packaged?, Ergonomics, Vol. 33, 1990, pp. 519–526.
173. McGuinness, D. L., Patel-Schneider, P. F., Usability Issues in Knowledge Representation Systems, Proceedings of the National Conference on Artificial Intelligence, 1998, pp. 608–614.
174. Meech, J. F., Thomas, P. J., Developing Usable Personal Systems: A Human Factors Perspective, IEE Colloquium (Digest), No. 140, 1995, pp. 2/1–2/3.
175. Miller, D. P., Automating Usability Metrics for Prototype Evaluation, Proceedings of the Human Factors Society Conference, 1990, pp. 222–224.
176. Miller, S., Jarrett, C., Setting Usability Requirements for a Web Site Containing a Form, Proceedings of the Society for Technical Communication Annual Conference, 2001, pp. 386–390.
177. Mirel, B., Critical Review of Experimental Research on the Usability of Hard Copy Documentation, IEEE Transactions on Professional Communications, Vol. 34, No. 2, 1991, pp. 109–122.
178. Miyoshi, T., Murata, A., Usability of Input Device Using Eye Tracker on Button Size, Distance Between Targets and Direction of Movement, Proceedings of the IEEE International Conference on Systems, Man, and Cybernetics, 2001, pp. 227–232.
179. Moeller, E. W., The Latest Web Trend: Usability?, Proceedings of the IEEE International Professional Communication Conference, 2001, pp. 151–158.
180. Moeller, E. W., Usability: A Web Trend for the Future? Proceedings of the Society for Technical Communication Annual Conference, 2002, pp. 291–296.
181. Mohageg, M., et al., User Interface for Accessing 3D Content on the World Wide Web, Proceedings of the Conference on Human Factors in Computing Systems, 1996, pp. 466–472.
182. Molich, R., Nielsen, J., Improving a Human-Computer Dialogue, Communications of the ACM, Vol.

33, No. 3, 1990, pp. 338–348.

183. Monk, A., Wright, P., Haber, J., Davenport, L., Improving Your Human-Computer Interface: A Practical Approach, Prentice-Hall, Inc., Englewood Cliffs, New Jersey, 1992.
184. Morgan, M. R. P., Crossing Disciplines: Usability as a Bridge Between System, Software,and Documentation, Technical Communication, Vol. 42, No. 2, 1995, pp. 303–306.
185. Morland, D. V., Human Factors Guidelines For Terminal Interface Design, Communications of the ACM, Vol. 26, No. 7, 1983, pp. 484–494.
186. Morris, M. G., Dillon, A. P., Importance of Usability in the Establishment of Organizational Software Standards for End User Computing, International Journal of Human -Computer Studies, Vol. 45, No. 2, 1996, pp. 243–258.
187. Mrazek, D., Rafeld, M., Integrating Human Factors On a Large Scale: "Product Usability Champions", Proceedings of the Conference on Human Factors in Computing Systems, 1992, pp. 565–570.
188. Mulkens, E., Roberts, J., Effects of Display Geometry and Pixel Structure on Stereo Display Usability, Proceedings of the International Society for Optical Engineering Conference, 2001, pp. 276–289.
189. Muller, M. J., Czerwinski, M., Organizing Usability Work to Fit the Full Product Range, Communications of the ACM, Vol. 42, No. 5, 1999, pp. 87–90.
190. Nagasawa, S., Fuzzy Kansei Evaluation of VCR's Usability, Proceedings of the IEEE International Conference on Systems, Man, and Cybernetics, 1999, pp. VI.290–VI.293.
191. Nardi, The Use of Ethnographic Methods in Design and Evaluation, in the Handbook of Human-Computer Interaction edited by M. G., Helander, T. K. Landauer, and P. V. Prabhu, North Holland Publishing Company, Amsterdam, 1997, pp. 361–367.
192. Neilsen, J., Designing Web Usability: The Practice of Simplicity, New Riders, Indianapolis, Indiana, 2000.
193. Nielsen J., Mack, R. L., Editors, Usability Inspection Methods, John Wiley and Sons, New York, 1994.
194. Nielsen, J., Faber, J. M., Improving System Usability Through Parallel Design, Computer, Vol. 29, No. 2, 1996, pp. 29–35.
195. Nielsen, J., Finding Usability Problems Through Heuristic Evaluation, Proceedings of the ACM Conference on Human Factors in Computing Systems, 1992, pp. 373–380.
196. Nielsen, J., Getting Usability Used, in Human-Computer Interaction, edited by K. Nordby, P. Helmersen, D. J., Gilmore, and S. Arnesen, Chapman and Hall, London, 1995, pp. 3–12.
197. Nielsen, J., Guerrilla HCI: Using Discount Usability Engineering to Penetrate the Intimidation Barrier, in Cost Justifying Usability edited by R. G. Bias and D. J. Mayhew, Morgan Kaufmann Publishers, San Francisco, 1994, pp. 245–272.
198. Nielsen, J., Iterative User-Interface Design, computer, Vol. 26, No. 11, 1993, pp. 32–41.
199. Nielsen, J., Landauser, T. K., Mathematical Model of the Finding of Usability Problems, Proceedings of the Conference on Human Factors in Computing Systems, 1993, pp. 206–213.
200. Nielsen, J., Levy, J., Measuring Usability Preference Vs. Performance, Communications of the ACM, Vol. 37, No. 4, 1994, pp. 66–75.
201. Nielsen, J., Molich, R., Teaching User Interface Design Based on Usability Engineering, ACM SIGCHI Bulletin, Vol. 21, No. 1, 1989, pp. 45–48.
202. Nielsen, J., The Usability Engineering Life Cycle, Computer, Vol. 25, 1992, pp. 12–22.
203. Nielsen, J., Usability Inspection Methods, Proceedings of the Conference on Human Factors in Computing Systems, 1995, pp. 377–378.
204. Nielsen, J., Usability Metrics: Tracking Interface Improvements, IEEE Software, Vol. 13, 1996, pp. 12–13.
205. Nielsen, J., What Do Users Really Want?, International Journal of Human-Computer Interaction, Vol. 1, No. 2, 1989, pp. 137–147.
206. Norman, D. A., Design Rules Based on Analyses of Human Error, Communications of the ACM, Vol. 26, No. 4, 1983, pp. 254–258.
207. Norman, D. A., The Design of Everyday Things, Doubleday-Currency, New York, 1990.
208. Obata, T., Daimon, T., Kawashima, H., Cognitive Study of In-Vehicle Navigation Systems: Applying Verbal Protocol Analysis to Usability Evaluation, Proceedings of the IEEE–IEE Vehicle Navigation

and Information Systems Conference, 1993, pp. 232–237.

209. Ogawa, K., Evaluation Method of Computer Usability Based on Human-to-Computer Information Transmission Model, Ergonomics, Vol. 35, No. 5–6, 1992, pp. 557–590.
210. Ohnemus, K. R., Incorporating Human Factors in the System Development Life Cycle: Marketing and Management Approaches, Proceedings of the IEEE International Professional Communication Conference, 1996, pp. 46–52.
211. Overbye, T., et al., Human Factors Analysis of Power System Visualizations, Proceedings of the Hawaii International Conference on System Sciences, 2001, pp. 45–47.
212. Palanque, P., Patern, F., Fields, B., Designing, User Interfaces for Safety Critical Systems, SIGCHI, Vol. 30, No. 4, 1998, pp. 8–12.
213. Paolini, P., Hypermedia, the Web and Usability Issues, Proceedings of the International Conference on Multimedia Computing & Systems, 1999, pp. 111–115.
214. Parkes, A. M., Case Study: A Safety and Usability Evaluation of Two Different Car Phone Designs, International Journal of Vehicle Design, Vol. 26, No. 1, 2001, pp. 12–29.
215. Patel, S. C., Drury, C. G., Prabhu, P., Design and Usability Evaluation of Work Control Documentation, Proceedings of the Human Factors and Ergonomics Society Conference, 1993, pp. 1156–1160.
216. Paterno. F., Santoro, C., Preventing User Errors by Systematic Analysis of Deviations From the System Task Model, International Journal of Human Computer Studies, Vol. 56, 2002, pp. 225–245.
217. Pearrow, M., Web Site Usability Handbook, Charles River Media, Rockland, Massachusetts, 2000.
218. Perlman, G., Achieving Universal Usability by Designing for Change, IEEE Internet Computing, Vol. 6, No. 2, 2002, pp. 46–55.
219. Potosnak, K. M., Koffler, R. P., Testing For Usability, Proceedings of the AFIPS Conference, 1987, pp. 77–84.
220. Potter, P., Usability Labs Make a Difference, Insurance Software Review, Vol. 14, No. 6, 1989, pp. 15–17.
221. Prail, A., Kahn, M. J., Usability Inspections:Their Potential Contribution, Designing for Diversity, Proceedings of the Human Factors and Ergonomics Society Conference, 1993, pp. 304–308.
222. Preece, J., Sociability and Usability in Online Communities: Determining and Measuring Success, Behavior and Information Technology, Vol. 20, No. 5, 2002, pp. 147–156.
223. Radle, K., Young, S., Partnering Usability with Development: How Three Organizations Succeeded, IEEE Software, Vol. 18, No. 1, 2001, pp. 38–45.
224. Ramey, J., Communication: Usability Engineering, Proceedings of the IEEE Conference on Engineering Communication (IPCC), 1991, pp. 53–57.
225. Ranch, T., Kahler, S., Flanagan, G., Usability Management Maturity, Part II: Usability Techniques—What Can You Do? SIGCHI, Vol. 28, No. 4, 1996, pp. 7–9.
226. Randolph, E., Ballard, L., In-house Usability Training: Culture Change You Can Afford, Proceedings of the Society for Technical Communication Annual Conference, 2001, pp. 33–36.
227. Ravden, S. J., Johnson, G. I., Evaluating Usability of Human-Computer Interfaces: A Practical Method, Ellis Horwood Publishers, Chichester, U.K., 1989.
228. Rideout, T. B., et al., Evolving the Software Usability Engineering Process at Hewlett-Packard, Proceedings of the IEEE International Conference on Systems, Man, and Cybernetics, 1989, pp. 229–234.
229. Riihiaho, S., The Pluralistic Usability Walk-Through Method, Ergonomics in Design, Vol. 10, No. 3, 2002, pp. 23–27.
230. Riley, C. A., McConkie, A. B., Designing for Usability: Human Factors in a Large Software Development Organization, Proceedings of the IEEE International Conference on System, Man, and Cybernetics, 1989, pp. 225–228.
231. Ritzel, K., Donelson, T. H., Human Factors Gets Cooking, Ergonomics in Design, Vol. 9, No. 1, 2001, pp. 15–19.
232. Robertson, J. W. , Usability and Children's Software: A User Centered Design Methodology, Journal of Computing in Childhood Education, Vol. 5, No. 3/4, 1994, pp. 257–271.
233. Rohn, J. A., The Usability Engineering Centers at Sun Microsystems, Behavior and Information Technology, Vol. 13, No. 1–2, 1994, pp. 25–35.
234. Rosenbaum, S., Ramey, J., Current Issues in Assessing and Improving Documentation Usability,

Proceedings of the Conference on Human Factors in Computing Systems, 1995, pp. 333–336.

235. Rosenbaum, S., Rohn, J. A., Humburg, J., Toolkit for Strategic Usability: Results from Workshops, Panels, and Surveys, Proceedings of the Conference on Human Factors in Computing Systems, 2000, pp. 337–344.
236. Rosson, M. B., Carroll, J. M., Usability Engineering: Scenario-Based Development of Human-Computer Interaction, Academic Press, San Francisco, 2002.
237. Rosson, M. B., Maass, S., Kellogg, W. A., The Designer as User: Building Requirements for Design Tools from Design Practice, Communications of the ACM, Vol. 31, No. 11, 1989, pp. 1288–1297.
238. Rubinstein, R., Hersh, H., The Human Factor: Designing Computer Systems for People, Digital Press, Maynard, Massachussetts, 1984.
239. Salminen, A., Tiitinen, P., Lyytikainen, V., Usability Evaluation of a Structured Document Archive, Proceedings of the Hawaii International Conference on System Sciences, 1999, p. 71–73.
240. Salvendy, G., Editor, Handbook of Human Factors and Ergonomics, John Wiley and Sons, New York, 1997.
241. Scerbo, M. W., Usability Engineering Approach to Software Quality, Annual Quality Congress Transactions, Vol. 45, 1991, pp. 726–733.
242. Schneider, M. F., Why Ergonomics Can No Longer Be Ignored, Office Administration and Automation, Vol. 46, No. 7, 1985, pp. 26–29.
243. Scholtz, J., Case Study: Developing a Remote Rapid and Automated Usability Testing Methodology for On-Line Books, Proceedings of the Hawaii International Conference On System Sciences, 1999, p. 69–70.
244. Scholtz, J., Shneiderman, B., Introduction to Special Issue on Usability Engineering, Emperical Software Engineering, Vol. 4, No. 1, 1999, pp. 5–10.
245. Schott, F., Olsen, M., Designing Usability in Systems: Driving for Normalcy, Datamation, Vol. 34, No. 10, 1988, pp. 5–10.
246. Schrier, J. R., Reducing Stress Associated with Participating in a Usability Test, Proceedings of the Human Factors Society Conference, 1993, pp. 1210–1214.
247. Schultz, E. E., et al., Usability and Security an Appraisal of Usability Issues in Information Security Methods, Computers and Security, Vol. 20, No. 7, 2001, pp. 620–634.
248. Sears, A., Jacko, J. A., Understanding the Relation Between Network Quality of Service and the Usability of Distributed Multimedia Documents, Human-Computer Interaction, Vol 15, No. 1, 2001, pp. 43–68.
249. Serafin, C., et al., Car Phone Usability: A Human Factors Laboratory Test, Proceedings of the Human Factors and Ergonomics Society Conference, 1993, pp. 220–224.
250. Shackel, B., IBM Makes Usability as Important as Important Functionality, The Computer Journal, Vol. 29, No. 3, 1986, pp. 10–14.
251. Shackel, B., Richardson, S., editors, Human Factors for Informatics Usability, Cambridge University Press, New York, 1991.
252. Shelden, S. , Vaughan, M., The Internet Usability Engineer, Ergonomics in Design, Vol. 9, 2001, pp. 27–28.
253. Shelden, S., Vaughan, M., The Internet Usability Engineer, Ergonomics in Design, Vol. 9, No. 2, 2001, pp. 27–28.
254. Shneiderman, B., Hochheiser, H., Universal Usability as a Stimulus to Advanced Interface Design, Behaviour and Information Technology, Vol. 20, No. 5, 2002, pp. 367–376.
255. Shroyer, R., Recruiting and Mentoring Usability Specialists, Proceedings of the Society for Technical Communication Annual Conference, 2001, pp. 273–277.
256. Sidhu, C. K., Coyle, G., Usability Engineering of Speech-Based Services, British Telecommunications Engineering, Vol. 14, 1996, pp. 337–340.
257. Silverman, D. R., Spiker, V. A., Usability Assessment of Virtual Realty Simulation for Aerial Gunner Training, Proceedings of the Human Factors and Ergonomics Society Conference, 1997, pp. 1218–1222.
258. Skelton, T. M., Testing the Usability of Usability Testing, Technical Communication, Vol. 39, No. 3, 1992, pp. 343–359.
259. Smelcer, J. B., User Errors in Database Query Composition, International Journal of Human Computer

Studies, Vol. 42, 1995, pp. 353–381.

260. Smilowitz, E. D., Darnell, M. J., Benson, A. E., Are We Overlooking Some Usability Testing Methods? A Comparison of Lab, Beta, and Forum Tests, Proceedings of the Human Factors and Ergonomics Society Conference, 1993, pp. 300–303.
261. Smith, S. L., Mosier, J. N., Design Guidelines for Designing User Interface Software, Report No. MTR-10090, The MITRE Corporation, Bedford, Massachusetts, 1986.
262. Spencer, R., The Streamlined Cognitive Walkthrough Method: Working Around Social Constraints Encountered in a Software Development Company, Proceedings of the Conference on Human Factors in Computing Systems, 2000, pp. 353–359.
263. Spencer, R. H., Computer Usability Testing and Evaluation, Prentice Hall, Englewood Cliffs, New Jersey, 1985.
264. Stanney, K. M., Mollaghasemi, M., Virtual Environment Usability Engineering, Proceedings of the Virtual Reality Annual International Symposium, 2002, pp. 305–307.
265. Stanyer, D., Procter, R., Improving Web Usability with the Link Lens, Computer Networks, Vol. 31, No. 11, 1999, pp. 1533–1544.
266. Suchman, L., Plans, and Situated Actions: The Problem of Human-Machine Communication, Cambridge University Press, Cambridge, U.K., 1987.
267. Sulaiman, S., Usability and the Software Production Life Cycle, Proceedings of the Conference on Human Factors in Computing Systems, 1996, pp. 61–62.
268. Sullivan, K., Horwitz, K. J., Windows 95 User Interface: A Case Study In Usability Engineering, Proceedings of the Conference on Human Factors in Computing Systems, 1996, pp. 473–480.
269. Sullivan, P., Usability in the Computer Industry: What Contribution Can Longitudinal Field Studies Make?, Proceedings of the International Conference on Professional Communication Conference, (IPCC'89), 1989, pp. 12–16.
270. Summer, T., The High-Tech Toolbelt: A Study of Designers in the Workplace, Proceedings of the Conference on Human Factors in Computing Systems, 1995, pp. 178–185.
271. Sy, D., Critical Factors for Starting Usability Engineering in the Workplace, Proceedings of the IEEE International Conference on Professional Communications, 1994, pp. 118–123.
272. Szczuer, M., Usability Testing on a Budget: A NASA Test Case Study, Behavior and Information, Vol. 13, No. 2, 1994, pp. 106–118.
273. Terazzi, A., Giordano, A., Minco, G., How Can Usability Measurement Affect the Reengineering Process of Clinical Software Procedures?, International Journal of Medical Informatics, Vol. 52, No. 1–3, 1998, pp. 229–234.
274. Theng, Y. L., Mohd-Nasir, N., Thimbleby, H., Purpose and Usability of Digital Libraries, Proceedings of the ACM International Conference on Digital Libraries, 2000, pp. 238–239.
275. Thovtrup, H., Nielsen, J., Assessing the Usability of a User Interface Standards, Proceedings of the Conference on Human Factors in Computing Systems, 1991, pp. 335–342.
276. Tighe, P. G., Usable Usability, Hewlett-Packard Journal, Vol. 47, 1996, pp. 88–93.
277. Traub, P., Optimising Human Factors Integration in System Design, Engineering Management Journal, Vol. 6, No. 2, 1996, pp. 93–98.
278. Trenner, L., Bawa, J., The Politics of Usability: A Practical Guide to Designing Usable Systems in Industry, Springer Verlag, New York, 1998.
279. Tyldesley, D. A., Employing Usability Engineering in Development of Office Products, Computer Journal, Vol. 31, No. 5, 1988, pp. 431–436.
280. Utter, S., Enhancing the Usability of Parallel Debuggers, Proceedings of the Hawaii International Conference on System Sciences, 1989, pp. 468–469.
281. Valero, M. A., et al, Usability engineering in Patients Follow-Up Medical Information Systems, Proceedings of the Annual IEEE International Conference on Engineering in Medicine and Biology, 1996, pp. 2307–2308.
282. Van Waes, L., Thinking Aloud as a Method for Testing the Usability of Websites! The Influence of Task Variation on the Evaluation of Hypertext, IEEE Transactions on Professional Communication, Vol. 43, No. 3, 2000, pp. 279–291.
283. Virzi, R. A.,Usability Inspection Methods, in the Handbook of Human-Computer Interaction, Edited by M. Helander, T. K. Landauer, and P. Prabhu, Elsevier Science Publishers, Amsterdam, 1997, pp.

231–252.

284. Virzi, R. A., Sokolov, J. L., Karis, D., Usability Problem Identification Using Both Low-and High-Fidelity Prototypes, Proceedings of the Conference on Human Factors in Computing Systems, 1996, pp. 236–243.
285. Vora, P., Classifying User Errors in Human Computer Interactive Tasks, Common Ground (Usability Professional Association), Vol. 5, No. 2, 1995, pp. 15–16.
286. Vredenburg, K., Increasing Ease of Use Communications of the ACM, Vol. 42, No. 5, 1999, pp. 67–71.
287. Wenger, M. J., Spyridakis, J. H., Relevance of Reliability and Validity to Usability Testing, IEEE Transactions on Professional Communications, Vol. 32, No. 4, 1989, pp. 265–271.
288. Wharton, C., Bradford, J., Jeffries, R., Franzke, M., Applying Cognitive Walkthroughs to More Complex User Interfaces: Experiences, Issues, and Recommendations, Proceedings of the ACM Conference on Human Factors in Computing Systems, 1992, pp. 381–388.
289. Whiteside, J., Bennett, J., Holzblatt, K., Usability Engineering: Our Experience and Evolution, in Handbook of Human-Computer Interaction, edited by M. Halander and K. Holzblatt, Elsevier Science, New York, 1988, pp. 791–817.
290. Wiklund, M. E., AAMI's Influence On the Design of User-friendly Medical Devices, Biomedical Instrumentation and Technology, Vol., 29, No. 3, 1995, pp. 229–231.
291. Wiklund, M. E., Editor, Usability in Practice: How Companies Develop User-Friendly Products, Academic Press, London, 1994.
292. Wiklund, M. E., Medical Device and Equipment Design: Usability Engineering and Ergonomics, Interpharm Press, Buffalo Grove, Illinois, 1995.
293. Wilkund, M. E., How to Implement Usability Engineering, Medical Device and Diagnostic Industry, Vol. 15, No. 9, 1993, pp. 68–73.
294. Wilson, C. E., Rosenbaum, S. L., Advanced Issues in Usability: A Progression, Proceedings of the Society for Technical Communication Annual Conference, 1996, pp. 290–293.
295. Wise, M., Bellaver, R., Usability Testing as a Teaching Tool, Ergonomics in Design, Vol. 15, No. 2, 1997, pp. 11–17.
296. Wixon, D., Ramey, J., editors, Field Methods Casebook for Software Design, John Wiley and Sons, New York, 1996.
297. Yates, R. F., Human Factors: An Overview, British Telecom Engineering, Vol. 7, Part 4, 1989, pp. 246–254.
298. Zhai, S., User Performance in Relation to 3D Input Device Design, Computer Graphic, Vol. 32, No. 4, 1998, pp. 50–54.
299. Zhang, Z., Basili, V., Shneiderman, B., Empirical Study of Perspective-Based Usability Inspection, Proceedings of the Human Factors Society Conference, 1998, pp. 1346–1350.
300. Zimmerman, D., Slater, M., Kandall, P., Risk Communication and Usability Case Study: Implications for Web Site Design, Proceedings of the IEEE International Conference on Professional Communication, 2001, pp. 445–452.

# INDEX